Trajectory Planning Using Dynamics and Power Models

This book shows how to plan trajectories (i.e. time-dependent paths) for autonomous robots using a dynamic model within the A* framework.

Drawing from optimal control's model predictive control framework, the book develops a paradigm called Sampling Based Model Predictive Optimization (SBMPO), which generates graph trees through input sampling of a dynamic model, enabling A*-type algorithms to find optimal trajectories. The book covers various robotic platforms and tasks, including manipulators lifting heavy loads, mobile robots navigating steep hills, energy-efficient skid-steered movements, thermally informed space exploration planning, and climbing robots in obstacle-rich environments. It also explores methods for updating dynamic models for robust operation and provides sample code for applying SBMPO to additional problems.

This resource is aimed at researchers, engineers, and advanced students in motion planning and control for robotic and autonomous systems.

Trajectory Planning Using Dynamics and Power Models

A Heuristics Based Approach

Camilo Ordonez, Mario Harper,
Jonathan T. Boylan, and
Emmanuel G. Collins, Jr.

CRC Press
Taylor & Francis Group
Boca Raton London New York

CRC Press is an imprint of the
Taylor & Francis Group, an **Informa** business

A CHAPMAN & HALL BOOK

First edition published 2026
by CRC Press
2385 NW Executive Center Drive, Suite 320, Boca Raton FL 33431

and by CRC Press
4 Park Square, Milton Park, Abingdon, Oxon, OX14 4RN

CRC Press is an imprint of Taylor & Francis Group, LLC

© 2026 Camilo Ordonez, Mario Harper, Jonathan T. Boylan, and Emmanuel G. Collins, Jr.

MATLAB® and Simulink® are trademarks of The MathWorks, Inc. and are used with permission. The MathWorks does not warrant the accuracy of the text or exercises in this book. This book's use or discussion of MATLAB® or Simulink® software or related products does not constitute endorsement or sponsorship by The MathWorks of a particular pedagogical approach or particular use of the MATLAB® and Simulink® software.

ISBN: 978-1-041-03440-7 (hbk)
ISBN: 978-1-041-03444-5 (pbk)
ISBN: 978-1-003-62383-0 (ebk)

DOI: 10.1201/9781003623830

Typeset in Latin Modern font
by KnowledgeWorks Global Ltd.

Publisher's note: This book has been prepared from camera-ready copy provided by the authors.

To my family.
Camilo Ordonez

To my family—your unwavering love, support,
and belief in me have made this journey
possible.
Mario Harper

Dedicated to my family, whose love and
support have been my greatest source of
strength and inspiration.
Jonathan Tyler Boylan

Contents

Preface

A* algorithms have been proven to be highly effective in path planning based on optimizing distance. In fact, they are the basis of path planning in both the Google Maps app and the Apple Maps app. The use of a reasonable heuristic or cost-to-go estimate, i.e., an estimate of the cost from a current configuration to the goal configuration, enables A*-type algorithms to compute very quickly, a requisite feature for real-time planning algorithms. This book provides research that extends the use of A*-type algorithms by providing an answer to this fundamental question: How does one plan trajectories (i.e., time-dependent paths) for an autonomous robot using a dynamic model within the A* framework?

To provide a solid answer to this question, the authors borrow from the model predictive control framework of optimal control theory and develop a paradigm called Sampling-Based Model Predictive Optimization, which uses input sampling of a dynamic model to generate graph trees for which an A*-type algorithm can be used to find the optimal trajectory. The book considers the following questions that arise in the development of the SBMPO framework:

- What type of optimization cost functions arise in practical applications?

- How does one develop heuristics for these cost functions?

- How does one handle dynamics that are too complex to directly integrate or propagate within a real-time algorithm?

- How does one account for the lower-level feedback controllers that are part of the robot dynamics?

To show that SBMPO is applicable to real-world trajectory planning problems, the book reports hardware-based experimental results for the following problems:

- Momentum-based planning for mobile robots through thick vegetation and steep hills;

- Momentum-based planning for manipulators lifting heavy loads;

- Energy-efficient motion planning for skid-steered robots;

- Distance-optimal planning for the TAILS Climbing Robot using an experimentally learned model.

Additionally, the book reports simulation results for other important applications, including minimum-time planning for autonomous spacecraft and thermally informed motion planning for mobile robots deployed for space exploration.

The authors know that there are many rich and important problems to be solved related to the SBMPO framework. Hence, the book provides directions for future work with a focus on the learning of dynamic models, the learning of heuristics, and the development of a multiple heuristic SBMPO framework.

Finally, it should be mentioned that the results of this book are focused on robots but are applicable in other optimization-based domains. In fact, one of the authors has worked with others to develop an adaptive control algorithm for nonlinear systems based on SBMPO and applied it to microjet-based flow separation control.

Author Biographies

Camilo Ordonez received a B.S. in Electronics Engineering from Pontificia Bolivariana University in 2003. He obtained his M.S. and Ph.D. degrees in Mechanical Engineering from Florida State University in 2006 and 2010, respectively. Currently, he is a faculty member in the department of mechanical engineering at the FAMU-FSU College of Engineering. He is part of the Center for Intelligent Systems, Controls, and Robotics (CISCOR) and the Energy and Sustainability Center. His research interests include dynamic modeling of legged and wheeled vehicles, terrain identification, and motion planning.

Mario Harper is a professor of Computer Science at Utah State University and the director of the Decision-making, Intelligence, Robotics, Electrification, and Transportation (DIRECT) Lab. With expertise spanning Artificial Intelligence, Machine Learning, Robotics, and Finance, Dr. Harper has contributed to many projects involving satellites, Mars rovers, military systems, and electrified transportation. His research integrates AI with electrification, space robotics, and intelligent systems, with a focus on practical applications in extreme environments. He received a B.S. in Physics and Economics from Utah State University, as well as an M.S. in Finance and Computational Science and a Ph.D. in Computer Science, both from Florida State University.

Jonathan Tyler Boylan earned a Bachelor's degree in Mechanical Engineering with a Minor in Computer Science from Florida State University in 2023. He is currently pursuing a Master's degree in Robotics at Florida State University, where he conducts research in the Scansorial and Terrestrial Robotics and Integrated Design (STRIDe) Lab at the FAMU-FSU College of Engineering. His work focuses on advancing decision-making algorithms for autonomous robotic systems, including autonomous ground vehicles (AGVs), quadrupedal robots, and other platforms. His research interests span dynamic modeling, motion planning, computer vision, and robotic control, aiming to bridge theoretical insights with practical innovations in autonomous robotics.

Emmanuel Collins currently serves as Dean of the J.B. Speed School of Engineering at the University of Louisville. He has had a long career as a researcher in the fields of controls and robotics. Upon graduating with his Ph.D. in Aeronautics and Astronautics from Purdue University, he was employed at Harris Corporation where he worked in the emerging field of flexible space structure control. He made major contributions to the development and demonstration of effective robust vibration control algorithms, culminating in an Honorary Superior Accomplishment Award from NASA. As a professor, he has made contributions to a variety of areas, including robust control, robust fault detection, proprioceptive terrain classification for robots, intelligence for both mobile robot planning and manipulator motion planning, and nonlinear adaptive control. He has always focused on developing or interpreting state-of-the-art optimization algorithms and applying them to real-world problems. This remains one of his passions.

CHAPTER 1

Introduction

In the realm of mobile robotics, traditional path planning is charged with finding collision-free paths from a start to a goal location or configuration from a purely geometric perspective [1]. On the other hand, trajectory planning seeks to find collision-free, time-dependent paths to take a mobile platform from a starting configuration to a desired goal. Traditional trajectory planning approaches were originally inspired from work on manipulators [1] and decompose the process into several stages: 1) a collision-free path (i.e., a set of way points) is built assuming that the vehicle is holonomic, 2) the path undergoes smoothing so that it is easier for the robot to follow, 3) a velocity profile is derived from the planned path, and 4) a feedback control law is employed to track the trajectory.

This traditional approach to motion planning is based upon geometric constraints and does not extend easily to incorporation of dynamic models. However, it is highly effective in many cases such as for highly maneuverable or slow-moving robots. In these problems, the difficulty in planning comes from the complexity of the robot's high-dimensional configuration space rather than from dynamics [2, 3]. Other types of motion that can be readily incorporated into geometric planning are those that include manifold-constraints [4], which involve constraining the local motion of a robot with constraints represented using algebraic equations instead of the dynamic equations used in kinodynamic planning, which is described in Section 1.2. However, for momentum-based or energy-efficient motion planning problems, incorporation of dynamic models is important and the traditional multi-stage approach to trajectory generation tends to be fragile as it is heavily dependent on the capacity of low level feedback control laws to "track" trajectories that are sometimes infeasible from a dynamic view point or simply energy inefficient.

1.1 EMERGENCE OF SAMPLING-BASED PATH PLANNING

A vast amount of work on path planning problems has dealt with deterministic planning algorithms that rely on different representations of the environment such as visibility graphs [5], Voronoi diagrams [6], Delauny triangulation [7], and cell decomposition methods [8]. Typically, once the environment is represented, heuristic-based search methods like A* [9] and its variants such as D* [10] and LPA* [11] are employed to find optimal paths. When the heuristics provide optimistic yet relatively accurate estimates of the cost to goal, these algorithms are fast and produce optimal results.

As the planning problems become more challenging, purely deterministic approaches have shown to be computationally demanding. To remedy this, sampling-based algorithms have emerged [12, 13]. These methods randomly or pseudo-randomly sample from the obstacle-free space to find a collision-free path to the goal. Different from the deterministic methods described above, sampling-based planners do not explicitly represent the environment but instead rely on a collision checking module [14].

The two most well-known sampling-based methods are Rapidly exploring random trees (RRTs) [12] and probabilistic road maps (PRMs) [15]. PRMs have two main stages: 1) a learning phase where a roadmap is built via sampling and 2) a query phase where solutions are found within the pre-existing roadmap. On the other hand, RRTs expand trees during the query phase and are more efficient for single query applications. There are multiple variants of RRTs that change the sampling techniques, the number of trees used to explore the free space, or the way in which the growth of the trees is biased. These methods are still widely regarded as suboptimal [13] and require post-processing stages to improve and smooth the generated paths.

More current work on sampling-based planners deals with the problem of adding optimality to the planned solutions. The most popular algorithm is RRT* [14]. However, this algorithm converges toward the optimal solution asymptotically, which means that the initial solutions are suboptimal. It has been shown that the convergence of RRT* toward the optimal solution can be slow, which motivated the development of Anytime algorithms [16, 17]. Geared toward real-time applications, anytime planners quickly return a solution to the planning query and improve on it if there is computational time remaining.

Variants of RRT* have considered traversability metrics for rovers moving on rough terrain [18], and have shown computational improvements for traversability-based RRT* over conventional RRT with traversability assessment.

1.2 KINODYNAMIC MOTION PLANNING

Robust planners move away from purely geometric planning into motion planning schemes that can more easily reflect the types of constraints to which the physical systems are subjected to [19]. That work [19] presented the first motion planning paradigm capable of obeying kinematic constraints (obstacles, non-traversable regions) and dynamic constraints (velocity bounds, acceleration, and force). This paradigm, formalized as kinodynamic motion planning, generated a minimum time trajectory for a point mass robot from a start position and velocity to a goal position and velocity while respecting velocity and acceleration constraints. A direct application of this algorithm to hardware was not practical because the employed robot model obeyed simple dynamics, was allowed to exert forces in any direction, and operated in continuous space. However, kinodynamic planning was made more practical using sampling-based techniques [20, 21].

Kinodynamic trajectory planning with more realistic vehicle models and consideration of rough terrain can be posed as a nonlinear programming problem that searches in the control space [22]. That methodology relies on numerical linearization of vehicle forward models and has been shown to be efficient and practical for real-time applications.

Motion primitives that create finite lattice representations of the state space have been used in combination with A* to find time optimal trajectories for differentially flat systems [23]. An extension [24] was proposed to allow quadrotors (differentially flat systems) to plan aggressive maneuvers through cluttered environments. That work [24] focused on the minimization of a cost function composed of a trade-off between control effort and time, which is suited for optimal control formulations.

Much work has been explored in kinodynamic planning leveraging the RRT* or PRM* algorithms [1, 25, 26]. RRT*-based kinodynamic algorithms typically struggle most due to needing a local planner that connects two states together, potentially slowing convergence, and varied suboptimal results due to the nature of their asymptotic optimality guarantees [14]. To combat these, many of these methods utilize heuristic functions which bias the search direction [27, 28] to speed up convergence inspired by A^* type algorithms [29]. Other strategies have examined direct propagation of controls [30] by sampling the input space to build a feasible trajectory. However, these methods still offer asymptotic optimality, which is a feature of the RRT-style algorithms. SBMPO eliminates the need for a connecting function by forward propagating the controls directly, allowing this technique to be used in situations where a solution to a Boundary Value Problem (BVP) is not available.

The Stable-Sparse RRT (SST*) [31] has many similar aspects to SBMPO. This algorithm strives to identify regions of exploration that are likely to have low costs and is able to maintain probabilistic completeness and asymptotic optimality while balancing an intelligent heuristic function. This method circumvents the need of a local connection planner by using a Monte Carlo-based forward propagation of the controls and reduces memory requirements by only retaining collision-free samples. While many of the difficulties of RRT* are alleviated, SST* does not consider non-distance optimization, and like RRT* methods, guarantees asymptotic or near-asymptotic optimality. Another point of note is that SST* requires both a nearest neighbor and k-nearest neighbor query when determining node removals, whereas SBMPO utilizes an implicit grid to prune the tree, simplifying the computations required per removal. It is however worth mentioning that there exist many robotics applications where optimal solutions are not required, and feasibility suffices. Examples of such motion planning methods are the non-optimal version of SST*, SST [32], and the KPIECE [33] method which, similar to SBMPO, is a forward propagation-based method.

In many cases, cost and heuristic functions focus heavily on finding optimal best-distance trajectories that are kinematically and dynamically feasible. However, many other cost functions have been considered in trajectory planning which optimize stealth [34, 35], time [36, 37], jerk [38, 39], and energy [40, 41]. Most of these cases rely on differentiability, holonomic platforms, or known inverse kinematics to generate a trajectory, all of which are constraints not required by SBMPO.

1.3 SAMPLING-BASED MODEL PREDICTIVE OPTIMIZATION

This book presents a planning methodology called Sampling-Based Model Predictive Optimization (SBMPO) that closes important gaps in the motion planning work by simultaneously incorporating a variety of features. First, it works with a variety of dynamic models of holonomic or non-holonomic robots; they can be linear or nonlinear, be derived from first principles or learned, and can also be non-invertible since SBMPO samples in the input (i.e., control) space, which alleviates the cumbersome need for local connection planners or inverse kinematics. Second, it enables a variety of optimality criteria to be used such as minimum distance, minimum time, or minimum energy; the latter relies on the use of power models that can be derived from dynamic models coupled with electrical models of the actuators. Third, it plans feasible trajectories directly, i.e., it doesn't rely on smoothing and post processing stages. Fourth, it uses a heuristically-guided A*-type algorithm that is capable of rapid computational convergence when the heuristic is sufficiently non-conservative. Fifth, it enables the planner to explicitly take into account the lower-level control system (e.g., that used to control the robot wheels). Finally, it has been experimentally demonstrated to work well in practice.

SBMPO has been applied to diverse problems in controls and robotics. In controls, it has been applied to control of flow separation [42], control of combustion processes [43], and control of microgrids [44]. In robotics, SBMPO has been used to plan time optimal trajectories for manipulators in the presence of heavy loads [45]. It has been employed to plan trajectories that exploit vehicle momentum to traverse mobility challenges such as steep hills and vegetation patches [46]. Also, SBMPO has been used [47] with detailed dynamic robot models to plan distance and energy optimal trajectories with non-holonomic robots. Furthermore, SBMPO has been used with legged robots [48, 49, 50], underwater vehicles [51], and spacecraft [52].

This book unifies and elaborates on previous publications and includes a more thorough and standardized discussion of SBMPO, more in-depth treatment about the usage of heuristics, provides guidelines to tune SBMPO parameters, and includes links to example problems with code and implementation details.

1.4 BOOK ORGANIZATION

The remainder of the book is structured as follows: Chapter 2 describes the SBMPO algorithm. Chapter 3 presents the heuristics required to directly generate efficient and near optimal trajectories. Chapter 4 illustrates case studies that demonstrate motion planning with dynamic models. Chapter 5 discusses learning of motion models and heuristics, and Chapter 6 provides concluding remarks of the book.

Bibliography

[1] S. M. LaValle, *Planning Algorithms*. Cambridge University Press, 2006.

[2] Z. Kingston, L. E. Kavraki, and M. Moll, "Sampling-based methods for motion planning with constraints," *Annual Review of Control, Robotics, and Autonomous Systems*, vol. 1, pp. 159–185, 2018.

[3] F. Islam, O. Salzman, and M. Likhachev, "Online, interactive user guidance for high-dimensional, constrained motion planning," *This is an arXiv preprint (arXiv:1710.03873)*, 2017.

[4] Z. Kingston, M. Moll, and L. E. Kavraki, "Exploring implicit spaces for constrained sampling-based planning," *The International Journal of Robotics Research*, vol. 38, no. 10-11, pp. 1151–1178, 2019.

[5] J. Latombe, *Robot Motion Planning*. Kluwer Academic Publishers, 1991.

[6] O. Takahashi and R. Schilling, "Motion planning in a plane using generalized Voronoi diagrams," *IEEE Transactions on robotics and automation 5.2 (1989):* vol. 5, 143–150.

[7] H. Yan, H. Wang, Y. Chen, and G. Gai, "Path planning based on constrained Delaunay triangulation," in *World Congress on Intelligent Control and Automation*, 6 2008.

[8] R. Brooks and T. Lozano-Perez, "A subdivision algorithm in configuration space for finding path with rotation," *IEEE Transactions on Systems, Man, and Cybernetics*, vol. 15, no. 2, pp. 224–233, 1985.

[9] P. Hart, N. Nilsson, and B. Raphael, "A formal basis for the heuristic determination of minimum cost paths," *IEEE Transactions on Systems, Man, and Cybernetics*, vol. 4, no. 2, pp. 100–107, 1968.

[10] A. Stentz, "Optimal and efficient path planning for unknown and dynamic environments," tech. rep., 1993.

[11] S. Koenig, M. Likhachev, and D. Furcy, "Lifelong planning A*," *Elsevier Science*, May 24 2005.

[12] S. LaValle, "Rapidly-exploring random trees: a new tool for path planning," tech. rep., 10 1998.

[13] M. Elbanhawi and M. Simic, "Sampling-based robot motion planning: A review," *IEEE Access*, vol. 2, pp. 56–77, 2014.

[14] S. Karaman and E. Frazzoli, "Sampling-based algorithms for optimal motion planning," *The International Journal of Robotics Research*, vol. 30, no. 7, pp. 846–894, 2011.

[15] L. Kavraki and J. Latombe, "Randomized preprocessing of configuration space for path planning," in *International Conference on Intelligent Robots*, vol. 3, pp. 1764–1771, 9 1994.

[16] S. Karaman, M. R. Walter, A. Perez, E. Frazzoli, and S. Teller, "Anytime motion planning using the RRT*," in *2011 IEEE International Conference on Robotics and Automation*, pp. 1478–1483, May 2011.

[17] R. Luna, I. A. Şucan, M. Moll, and L. E. Kavraki, "Anytime solution optimization for sampling-based motion planning," in *2013 IEEE International Conference on Robotics and Automation*, pp. 5068–5074, May 2013.

[18] R. Takemura and G. Ishigami, "Traversability-based RRT* for planetary rover path planning in rough terrain with lidar point cloud data," *Journal of Robotics and Mechatronics*, vol. 29, no. 5, pp. 838–846, 2017.

[19] B. Donald, P. Xavier, J. Canny, and J. Reif, "Kinodynamic motion planning," *Journal of the ACM (JACM)*, vol. 40, no. 5, pp. 1048–1066, 1993.

[20] L. E. Kavraki, P. Svestka, J.-C. Latombe, and M. H. Overmars, "Probabilistic roadmaps for path planning in high-dimensional configuration spaces," *IEEE Transactions on Robotics and Automation*, vol. 12, no. 4, pp. 566–580, 1996.

[21] S. M. LaValle and J. J. Kuffner Jr, "Randomized kinodynamic planning," *The International Journal of Robotics Research*, vol. 20, no. 5, pp. 378–400, 2001.

[22] T. M. Howard and A. Kelly, "Optimal rough terrain trajectory generation for wheeled mobile robots," *The International Journal of Robotics Research*, vol. 26, no. 2, pp. 141–166, 2007.

[23] S. Liu, N. Atanasov, K. Mohta, and V. Kumar, "Search-based motion planning for quadrotors using linear quadratic minimum time control," in *IEEE/RSJ International Conference on Intelligent Robots and Systems (IROS)*, pp. 2872–2879, IEEE, 2017.

[24] S. Liu, K. Mohta, N. Atanasov, and V. Kumar, "Search-based motion planning for aggressive flight in SE(3)," *IEEE Robotics and Automation Letters*, vol. 3, no. 3, pp. 2439–2446, 2018.

[25] B. Sakcak, L. Bascetta, G. Ferretti, and M. Prandini, "Sampling-based optimal kinodynamic planning with motion primitives," *Autonomous Robots*, vol. 43, no. 7, pp. 1715–1732, 2019.

[26] Z. Littlefield, D. Surovik, W. Wang, and K. E. Bekris, "From quasi-static to kinodynamic planning for spherical tensegrity locomotion," in *Robotics Research*, pp. 947–966, Springer, 2020.

[27] K. Bekris and L. Kavraki, "Informed and probabilistically complete search for motion planning under differential constraints," in *First International Symposium on Search Techniques in Artificial Intelligence and Robotics (STAIR), Chicago, IL*, 2008.

[28] Y. Li, Z. Littlefield, and K. E. Bekris, "Sparse methods for efficient asymptotically optimal kinodynamic planning," in *Algorithmic Foundations of Robotics XI*, pp. 263–282, Springer, 2015.

[29] C. Urmson and R. Simmons, "Approaches for heuristically biasing RRT growth," in *IEEE/RSJ International Conference on Intelligent Robots and Systems*, vol. 2, pp. 1178–1183, IEEE, 2003.

[30] G. Papadopoulos, H. Kurniawati, and N. M. Patrikalakis, "Analysis of asymptotically optimal sampling-based motion planning algorithms for Lipschitz continuous dynamical systems," *arXiv preprint arXiv:1405.2872*, 2014.

[31] Z. Littlefield and K. E. Bekris, "Efficient and asymptotically optimal kinodynamic motion planning via dominance-informed regions," in *IEEE/RSJ International Conference on Intelligent Robots and Systems (IROS)*, pp. 1–9, IEEE, 2018.

[32] Y. Li, Z. Littlefield, and K. E. Bekris, "Asymptotically optimal sampling-based kinodynamic planning," *The International Journal of Robotics Research*, vol. 35, no. 5, pp. 528–564, 2016.

[33] I. A. Sucan and L. E. Kavraki, "A sampling-based tree planner for systems with complex dynamics," *IEEE Transactions on Robotics*, vol. 28, no. 1, pp. 116–131, 2011.

[34] Z. Zhao, Y. Niu, Z. Ma, and X. Ji, "A fast stealth trajectory planning algorithm for stealth UAV to fly in multi-radar network," in *IEEE International Conference on Real-time Computing and Robotics (RCAR)*, pp. 549–554, IEEE, 2016.

[35] P. T. Kabamba, S. M. Meerkov, and F. H. Zeitz III, "Optimal path planning for unmanned combat aerial vehicles to defeat radar tracking," *Journal of Guidance, Control, and Dynamics*, vol. 29, no. 2, pp. 279–288, 2006.

[36] P. Huang and Y. Xu, "PSO-based time-optimal trajectory planning for space robot with dynamic constraints," in *IEEE International Conference on Robotics and Biomimetics*, pp. 1402–1407, IEEE, 2006.

[37] D. Constantinescu and E. A. Croft, "Smooth and time-optimal trajectory planning for industrial manipulators along specified paths," *Journal of Robotic Systems*, vol. 17, no. 5, pp. 233–249, 2000.

[38] A. Gasparetto and V. Zanotto, "A technique for time-jerk optimal planning of robot trajectories," *Robotics and Computer-Integrated Manufacturing*, vol. 24, no. 3, pp. 415–426, 2008.

[39] J. Huang, P. Hu, K. Wu, and M. Zeng, "Optimal time-jerk trajectory planning for industrial robots," *Mechanism and Machine Theory*, vol. 121, pp. 530–544, 2018.

[40] S. F. Saramago and V. S. Junior, "Optimal trajectory planning of robot manipulators in the presence of moving obstacles," *Mechanism and Machine Theory*, vol. 35, no. 8, pp. 1079–1094, 2000.

[41] P. Tokekar, N. Karnad, and V. Isler, "Energy-optimal trajectory planning for car-like robots," *Autonomous Robots*, vol. 37, no. 3, pp. 279–300, 2014.

[42] B. Reese, E. Collins, Jr., and F. Alvi, "A nonlinear adaptive method for microjet-based flow separation control,," *AIAA Journal*, vol. 54, pp. 3002–3014, 2016.

[43] B. Reese and E. Collins, "Sampling based control of a combustion process using a neural network model," in *IEEE International Conference on Systems, Man, and Cybernetics (SMC)*, vol. 1, pp. 966–972, 2014.

[44] N. Gupta, G. Francis, J. Ospina, A. Newaz, E. G. Collins, O. Faruque, R. Meeker, and M. Harper, "Cost optimal control of microgrids having solar power and energy storage," in *IEEE/PES Transmission and Distribution Conference and Exposition (T&D)*, pp. 1–9, IEEE, 2018.

[45] O. Chuy, E. Collins, A. Sharma, and R. Kopinsky, "Using dynamics to consider torque constraints in manipulators planning with heavy loads," *ASME Journal of Dynamic Systems, Measurement, and Control*, 2016.

[46] C. Ordonez, N. Gupta, O. Chuy, and E. Collins, "Momentum based traversal of mobility challenges for autonomous ground vehicles," in *Proceedings of the IEEE Conference on Robotics and Automation*, (Karlsruhe, Germany), pp. 752–759, 2013.

[47] N. Gupta, C. Ordonez, and E. G. Collins, "Dynamically feasible, energy efficient motion planning for skid-steered vehicles," *Autonomous Robots*, vol. 41, pp. 453–471, Feb 2017.

[48] M. Harper, J. Nicholson, E. Collins, J. Pusey, and J. Clark, "Energy efficient navigation for running legged robots," in *International Conference on Robotics and Automation (ICRA)*, pp. 6770–6776, 2019.

[49] C. Ordonez, N. Gupta, E. G. Collins, J. E. Clark, and A. M. Johnson, "Power modeling of the XRL hexapedal robot and its application to energy efficient motion planning," *Adaptive Mobile Robotics*, p. 689–696, 2012.

[50] C. Ordonez, J. Ordonez, J. Boylan, D. Vazquez, and J. Clark, "Thermally informed motion planning to enhance mission endurance of mobile robots," in *Proceedings of the 8th Thermal and Fluids Engineering Conference*, (College Park, MD), March 2023.

[51] C. V. Caldwell, D. D. Dunlap, and E. G. Collins, Jr., "Application of sampling based model predictive control to an autonomous underwater vehicle," *Ship Science and Technology*, vol. 4, pp. 55–63, July 2010.

[52] G. Francis, E. Collins, O. Chuy, and A. Sharma, "Sampling-based trajectory generation for autonomous spacecraft rendezvous and docking," in *AIAA Guidance, Navigation, and Control Conference*, (Boston, MA), 2013.

CHAPTER **2**

Review of SBMPO

This Chapter presents an overview of SBMPO. Section 2.1 defines the propagation model that is fundamental to SBMPO. Section 2.2 describes the input sampling used in SBMPO, including how an implicit grid is used to keep the size of the graph small. Section 2.3 describes the spatial and time-dependent meaning of a tree in the context of SBMPO. Section 2.4 explains Sampling-Based Model Predictive Control using the concept of receding horizon. Section 2.5 describes the major components of the SBMPO Algorithm and presents its key steps. Section 2.6 presents soundness and completeness results for the SBMPO Algorithm. Section 2.7 provides an illustration of the search process followed by SBMPO. Section 2.8 discusses application of SBMPO to high-dimensional systems. Section 2.9 provides insights into how to choose the grid resolution and sample period. Finally, Section 2.10 presents a baseline comparison of SBMPO to RRT*.

2.1 PROPAGATION MODEL

SBMPO assumes a *propagation model*, which is a nonlinear discrete-time model with the input and output shown in Fig. 2.1, wherein the figure and throughout this book we use the standard notation that the variable k is shorthand for the time kT, where T corresponds to both the sample and control update period.

DOI: 10.1201/9781003623830-2

As an illustration, the propagation model can be a discrete-time state-space model,

$$x(k+1) = f[x(k), u(k)], \qquad (2.1)$$
$$y(k+1) = h[x(k+1)]. \qquad (2.2)$$

However, the model may also be a neural network, simulation, input-output model, or any other model with the input and output as in Fig. 2.1. Illustrations of SBMPO's use with neural networks are shown in [1] and [2].

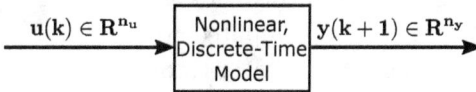

$$\mathbf{u(k)} \in \mathbf{R}^{n_u} \longrightarrow \boxed{\begin{array}{c} \text{Nonlinear,} \\ \text{Discrete-Time} \\ \text{Model} \end{array}} \xrightarrow{\ \mathbf{y(k+1)} \in \mathbf{R}^{n_y}\ }$$

Figure 2.1 SBMPO propagation model. SBMPO propagates a nonlinear, discrete-time model with the input and output shown here.

Though the propagation model of (2.1)–(2.2) and Fig. 2.1 can be complex, for two major reasons related to computational speed, the current implementations of SBMPO have used simple models in SBMPO such as the double integrator model, first introduced in Fig. 3.3 and (3.5)–(3.6) of Chapter 3. Both reasons are related to computational speed. First, while it is possible to develop heuristics for simple models such as the heuristics developed in Chapter 3 for the double integrator model, it is difficult to analytically develop heuristics for complex dynamics; however, it may be possible to learn heuristics for complex dynamics as discussed in Section 5.2. Also, practically propagating these models in real-time may be too computationally costly.

The double integrator model is used in Section 4.1 *Momentum-Based Planning* and appears in Section 4.1.3 *Double Integrator Model* as (4.3)–(4.4) and (4.5); it also appears in Fig. 4.4. However, to take into account torque or force constraints, the full dynamic model is used in the context of a computed torque controller as discussed in Section 4.1.4

Input Constraints. Table 2.1 summarizes the models used in this book for various planning problems.

Table 2.1 shows that in the planning problems considered in this book, integrator models, single or double, are used in most cases as the propagation model. The real exception is the use of the full dynamic model in minimum-time planning for autonomous spacecraft. This was possible largely because of the simplicity of that model. Table 2.1 also shows that in each of the considered planning problems the full dynamic model was always used, in most cases indirectly. The indirect use of the full dynamic model in this book is usually in conjunction with its appearance in a computed torque controller; in this case, it is used to enforce actuator torque constraints. *A subtle but important contribution of this book is to highlight ways in which full dynamic models can be used in planning without using these complex models as the propagation model.*

2.2 INPUT-SAMPLING

At each time instant k, SBMPO spatially discretizes the problem by sampling the input to the model in the region of feasible inputs. Although this discretization could be based on gridding, it is typically based on Halton sampling [3, 4, 5], as illustrated in Fig. 2.2(a), which guarantees low discrepancy. Random sampling can also be used if sampling occurs in high dimensions. The general advantage of sampling over gridding is that in gridding the number of points increases by some multiple of 2; for example, in the two-dimensional case of Fig. 2.2(b), the number of points increases as 2, 2^2, 2^3, ... However, sampling enables the number of points to increase by smaller increments (even an increment of 1) while maintaining a relatively even distribution in the space similar to gridding. This is highly advantageous in limiting the dimensionality of the problem.

Output sampling seems more natural to many in the planning community due to its ability to enable the output space to be explored uniformly. Hence, a pertinent question is why

Table 2.1 Summary of Planning Problems, Propagation Models, and Dynamics Models

Planning Problem	Section	Propagation Model	Use of Full Dynamics Model
General Momentum-Based Planning	4.1	Double Integrator (4.5)	Equations (4.8)–(4.10) used in Step 3 of Section 2.5
Momentum-Based Planning for Mobile Robots Through Mobility Challenges	4.1.5	Double Integrator (4.21)	Equations (4.19)–(4.20) used in Step 3 of Section 2.5
Momentum-Based Planning for Manipulators	4.1.6	Double Integrator (4.5)	Equations (4.8)–(4.10) used in Step 3 of Section 2.5
Minimum-Time Planning for Autonomous Spacecraft	4.2	Dynamic Model (4.29)	The dynamic model is used as the propagation model.
Energy Efficient Planning for Skid-Steered Vehicles	4.3	Single Integrator Model (4.41)	Planning uses the steady-state dynamic model, given by the Torque vs. Radius curves of Fig. 4.16; this model is used in Fig. 4.18 to compute the energetic cost $E(kT)$.
Thermally Informed Motion Planning for Legged Robots	4.4	Single Integrator Model (4.57)	The thermomechanical dynamic model (4.51) is used to compute the cost.
Energy Optimal Planning for LLAMA Legged Robot	5.1.2	Neural Network + Single Integrator, (5.7)–(5.9)	The dynamic model of Fig. 5.8 is used as the propagation model.

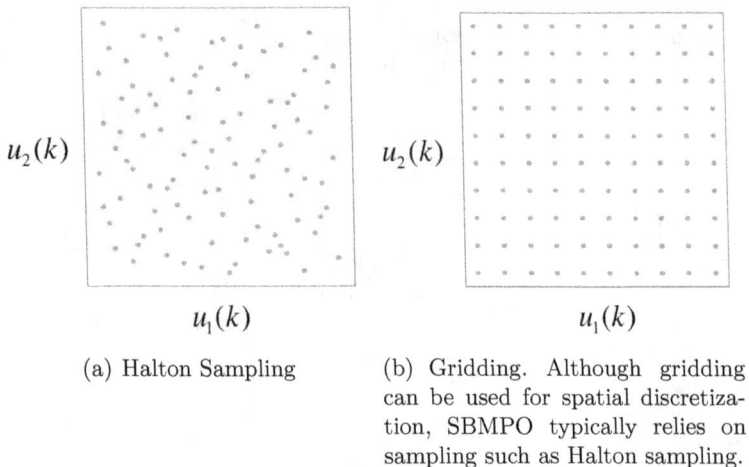

$u_2(k)$	$u_2(k)$
$u_1(k)$	$u_1(k)$
(a) Halton Sampling	(b) Gridding. Although gridding can be used for spatial discretization, SBMPO typically relies on sampling such as Halton sampling.

Figure 2.2 Input Sampling Techniques.

does SBMPO rely on input sampling? To answer this question refer again to Fig. 2.1, which shows the dimensions of the input $u(k)$ and output $y(k+1)$ are respectively n_u and n_y. Also, recognize that for many problems there are constraints on the inputs that must be enforced. The three primary reasons that input sampling is used are:

1. *Dimensionality Reduction.* If $n_u < n_y$, input sampling reduces the dimensionality of the problem.

2. *Non-Invertibility.* It may be difficult or impossible to determine $u(k)$ and its possible violation of the input constraints based on knowledge of $y(k + 1)$. This occurs when the dynamic model is complex or if $n_u \neq n_y$, especially if $n_u < n_y$.

3. *Violation of Input Constraints.* Even when the model is invertible, sampling the output may sometimes or even frequently lead to inputs that violate the input constraints.

However, if the propagation model is invertible, then output sampling can be used with SBMPO since the outputs can be sampled and the model inverted to produce the inputs.

2.2.1 Implicit Grid

It is possible that input sampling will produce some outputs that are closely spaced and can be practically treated as the same output. Hence, as illustrated in Fig. 2.3, to enforce spacing between the outputs, SBMPO uses an *implicit* grid [6]. In Fig. 2.3 the implicit grid recognizes that nodes v_2 and v_3 are close enough to be considered the same and updates the path to their grid cell to be the path corresponding to the edge cost c if $c < a + b$.

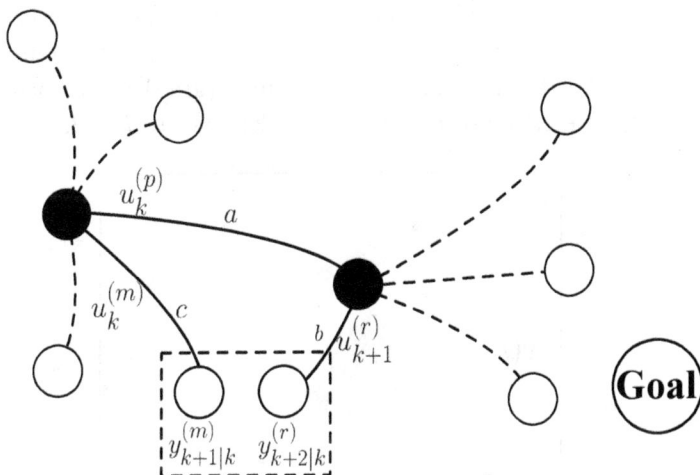

Figure 2.3 Implicit gridding. SBMPO uses an implicit grid to trim the graph. If a pair of nodes are sufficiently close, then one of them is eliminated. The variables a, b, and c represent edge costs.

The implicit grid is used to limit the size of the graph. Hence, the resolution of the grid is a significant factor in determining the performance of SBMPO. In general, the finer the grid the greater the algorithm computation time, due to the increased number of nodes in the graph; however, this also leads to a more optimal solution. It follows that the grid resolution r_g is a useful tuning tool for enabling SBMPO to effectively trade-off between solution quality and computational performance.

2.3 TREE GRAPH

At time k, SBMPO propagates each of the sampled inputs $u(k)$ through the propagation model of Fig. 2.1 to determine the corresponding output $y(k+1)$. This results in a tree as shown in Fig. 2.4. In this figure, $u_i^{(j)}$ denotes the jth sample at time i and $y(i|k)$ denotes a predicted output at time $i > k$. The notional SBMPO tree of Fig. 2.4 represents both spatial discretization (denoted by the branches) and time discretization denoted by the fact that each layer of nodes corresponds to a discrete time instance, e.g., the first (gray) layer corresponds to $k+1$, the second (dark black) layer to $k+2$, etc.

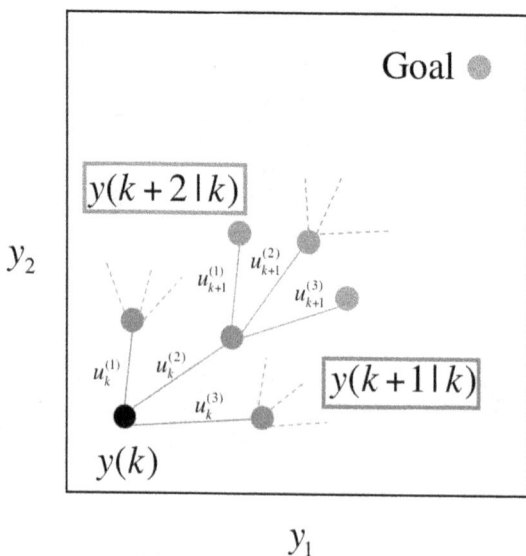

Figure 2.4 Illustrative SBMPO tree. This tree assumes the algorithm is optimizing at time k.

In Fig. 2.4, the (black) node corresponding to $y(k)$ is the parent node of the (gray) nodes corresponding to $y(k + 1|k)$, which are its children, and each gray node is a parent to a set of (gray) nodes corresponding to $y(k + 2|k)$, which are its children. Each child is connected to a parent by an *edge*, i.e, a branch. In SBMPO, each node is assigned a cost using the format of an A* algorithm [7, 8]. In particular, the cost f_{parent} of a parent node is given by

$$f_{parent} = g_{parent} + h_{parent}, \qquad (2.3)$$

where g_{parent} is the cost from the start node to the parent node and h_{parent}, the *heuristic*, is an estimate of the cost from the parent node to the goal. Determination of fairly accurate heuristics is important to the achievement of computationally fast trajectory generation and is considered in detail in Chapter 3. The cost f_{child} of a child node is given by

$$f_{child} = g_{child} + h_{child}, \qquad (2.4)$$

where h_{child} is the child's heuristic,

$$g_{child} = g_{parent} + edge\ cost, \qquad (2.5)$$

and *edge cost* is a cost assigned to the edge based on the underlying cost function that is being optimized such as distance, time, or energy.

SBMPO solves the optimization problem,

$$\min_{\{u_k, \cdots, u_{k+N-1}\}} J, \qquad (2.6)$$

where examples of the cost function J are given in Table 2.2. Note that the minimum deviation cost function is typical of a model predictive control cost function [9]. Hence, SBMPO is capable of both planning and model predictive control.

2.4   SAMPLING-BASED MODEL PREDICTIVE CONTROL

Sampling-Based Model Predictive Control (SBMPC) is SBMPO used within a receding horizon framework, such as

Table 2.2 Typical Cost Functions Optimized by SBMPO

Optimization Cost Function		Variable Definition
Minimum Distance	$\sum_{i=0}^{N-1} d_{k+i}$	d_j: distance from node j to $j+1$
Minimum Time	$\sum_{i=0}^{N-1} t_{k+i}$	t_j: time from node j to $j+1$
Minimum Energy	$\sum_{i=0}^{N-1} E_{k+i}$	E_j: energy from node j to $j+1$
Minimum Deviation	$\sum_{i=0}^{N-1} \|r_{k+i+1} - y_{k+i+1}\|^2$	r_j: reference input at time j; y_j: output at time j

that shown in Fig. 2.5. SBMPO is used to solve the optimization problem for each row, while SBMPC is considered the entire optimization process.

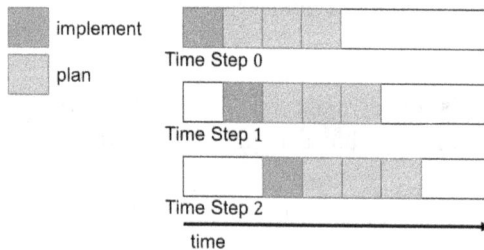

Figure 2.5 Illustration of receding horizon. SBMPO is sometimes used within this framework, which results in SBMPC.

2.5   SBMPO ALGORITHM

Fig. 2.6 illustrates the major components of SBMPO and their relationships. In this figure the initialization involves choosing the *branchout*, which is the number of input samples used at each time instant, the prediction horizon, and the implicit grid used for selected output variables. The output of the algorithm

```
┌─────────────────────────────────────┐
│          Initialization:            │
│  branchout, horizons, implicit grid │
└─────────────────────────────────────┘
                  │
                  ▼
┌─────────────────────────────────────┐
│          Node Selection:            │◄──────────┐
│     select node with highest priority          │
│   heuristic-based search (e.g., A* or LPA*)     │
└─────────────────────────────────────┘           │
                  │                                │
                  ▼                                │
┌─────────────────────────────────────┐           │
│          Input Sampling:            │           │
│ sample feasible control inputs using fixed, Halton, or │
│            random samples           │           │
└─────────────────────────────────────┘           │
                  │              ┌─────────────────┴──────┐
                  ▼              │       Repeat:          │
┌──────────────────────────────────────┐│ see stopping criteria │
│        Model Propagation:            ││      details          │
│   find next state via forward simulation ││ (step 6, Section 2.5) │
│(e.g., single (Eqs. 4.42,4.58) or double integrator (Eqs. 4.6, 4.22),└────────────────────────┘
│ dynamic model (Eq. 4.30), neural network (Fig. 5.8), etc) │
└──────────────────────────────────────┘           │
                  │                                │
                  ▼                                │
┌─────────────────────────────────────┐           │
│        Enforce Constraints:         │           │
│  e.g., state constraints, collisions, etc. │      │
└─────────────────────────────────────┘           │
                  │                                │
                  ▼                                │
┌─────────────────────────────────────┐           │
│        Add Node to Graph:           │───────────┘
│ cost node = cost of parent + edge cost + heuristic │
└─────────────────────────────────────┘
```

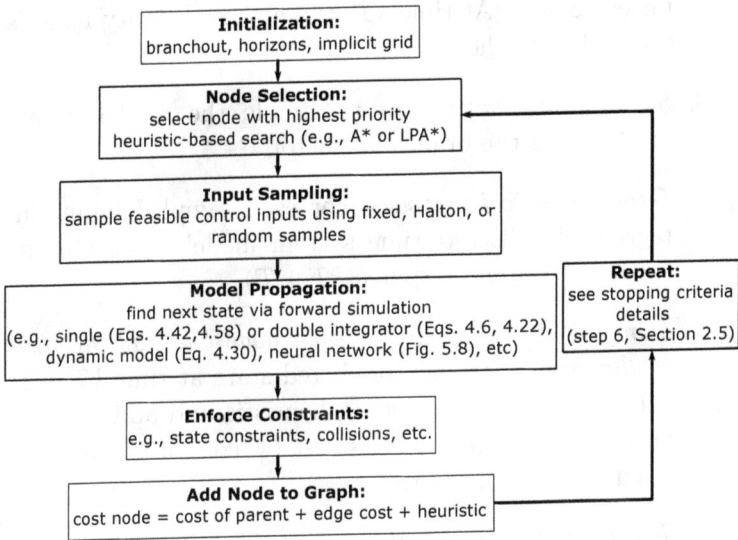

Figure 2.6 Major components of SBMPO.

is the optimal input trajectory and the corresponding output trajectory. The SBMPO paradigm allows the use of various heuristics-based graph search algorithms such as A* [7, 8], LPA* [10], and MHA* [11]. Sometimes it is prudent to refer to a particular implementation of SBMPO using the graph search algorithm in the name, e.g., SBMPO/LPA*. It is of interest to note that if the propagation model is chosen as

$$y(k + 1) = y(k) + u(k), \qquad (2.7)$$

and $u(k)$ is sampled in a grid pattern, then SBMPO/A* is equivalent to the standard grid-based A* that plans in the output space.

The Key Steps of SBMPO

1. *Initialization.* Let $k = 0$ and choose branchout factor B, prediction horizon N, control horizon M, sample period T, and implicit grid resolution r_g.

2. *Select a node with highest priority.* The nodes are collected in a queue in which they are arranged by priority

for expansion. At time kT, the highest priority node is selected from the queue.

3. *Sample the control space.* Sample the model inputs $u(kT)$ using the branchout factor B.

4. *Generate neighbor nodes.* For each sampled input, integrate the discrete-time system model to obtain the predicted system output $y(kT + T)$.

5. *Compute the heuristic for each node and add the node to the graph.* For each predicted state at time $kT + T$, calculate the corresponding heuristic and add the new node to the graph with its cost (= cost of parent node + edge cost + heuristic).

6. *Repeat 2 through 5 until one of the following stopping criteria is true.*

   (a) the goal is reached (suboptimal solution),

   (b) the goal is reached and there are no possible nodes in the priority queue that can improve the cost (optimal is reached),

   (c) the prediction horizon NT is reached (suboptimal),

   (d) the prediction horizon is reached and there are no possible nodes in the priority queue, or

   (e) the maximum number of allowable iterations is achieved.

The stopping criterion chosen by the user is dependent on the application.

2.6 ALGORITHM COMPLETENESS

Theorem 1 [2] (Soundness and Completeness) Assume $N = \infty$, an implicit grid is not used and the SBMPO algorithm is terminated using termination criteria 6(b). Then, upon termination the SBMPO algorithm will produce a sequence of nodes representing a minimal cost trajectory among those represented by the graph.

Note that Theorem 1 is based on the assumptions $N = \infty$, the absence of the use of an implicit grid, and the most rigorous termination criteria of the SBMPO algorithm, i.e., Step 6(b) of Section 2.5. All practical implementations of motion planning algorithms require computational speed. To achieve this, in practice SBMPO is implemented using finite N, an implicit grid to limit the size of the priority queue, and one of the other termination criteria in Section 2.5. Theorem 1 implies that if for a given implementation, N is sufficiently large, the implicit grid dimension resolution is sufficiently small, and the termination criteria given by Step 6(b) of Section 2.5 is used, SBMPO will find the optimal solution. Hence, the theorem shows that SBMPO is fundamentally based upon sound theory. However, optimal solutions will usually not be reached due to the need for computational speed. It is unclear at this time whether the restrictions upon which Theorem 1 is based may be relaxed and optimality still ensured. This is a subject for future research.

2.7 ILLUSTRATION OF SBMPO

To illustrate how SBMPO looks in operation, refer to Fig. 2.7, which shows a series of frames illustrating the search process followed by SBMPO.

2.8 APPLICATION OF SBMPO TO HIGH DIMENSIONAL SYSTEMS WITH DYNAMIC CONSTRAINTS

A relevant question is whether SBMPO can be applied to high-dimensional systems that have dynamic constraints. The answer to this question is *yes* in principle but may be *no* when computational speed is considered. Much depends on the complexity of the dynamic constraints associated with a given application of SBMPO. In this book, the highest

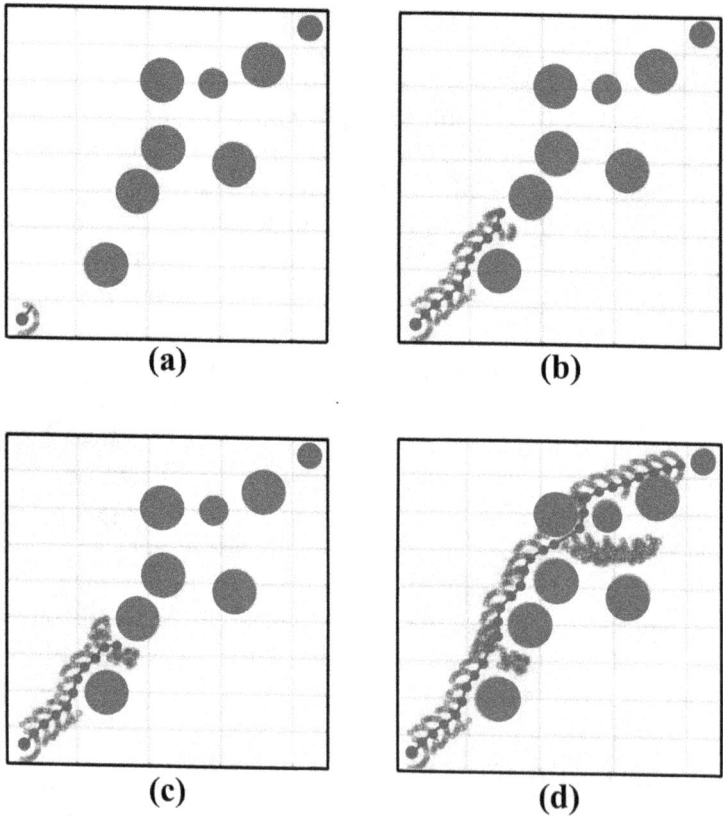

Figure 2.7 Sequence of frames (a)–(d) illustrating the search process followed by SBMPO to generate a minimum distance trajectory from the start (black circle) to the goal (filled gray circle). The black line denotes the SBMPO solution path and the nodes explored by SBMPO during the sampling process are represented in gray. Notice how there are some nodes in frames (c) and (d) that at some point in the exploration process were considered promising (i.e, they had a low overall f value (2.4) but eventually were rejected due to the appearance of new obstacles or the discovery of better nodes during the sampling process.

dimensional implementation of SBMPO is the 6 degrees of freedom (DOF) (12-D state space) autonomous spacecraft of Section 4.2, which considers planning for terminal velocities of the 6 states. Early SBMPO work demonstrated planning for 4 DOF of the Barrett Whole Arm Manipulator [12]; again, the main constraint was on the terminal velocities of the 4 joint angles. These are the highest dimensional systems considered to date. In contrast, in Section 4.1, which considers momentum-based planning such as climbing a steep hill or lifting a heavy object, only 1- or 2-dimensional systems are considered. If the hill is sufficiently steep or the object sufficiently heavy, then the feasible solution space may be small due to the torque limitations of the actuators and it would be difficult to practically apply SBMPO to high-dimensional momentum-based planning.

More experience is needed to better understand the dimensional limits of problems to which SBMPO is applied. However, two facts should be noted. First, as mentioned in Section 2.2, input sampling can help with the dimensionality problem. In particular, if the dimension of the system inputs is less than that of the system outputs, input sampling reduces the dimensionality of the problem. Second, if dynamic constraints are not part of the planning problem, such that the problem reduces to path planning, then SBMPO planning becomes A* planning and has the same benefits and limitations.

2.9 EFFECT OF GRID RESOLUTION AND SAMPLE PERIOD

In this section, we use the unicycle steering model to better understand the effects of the grid resolution g_r (as introduced in Section 2.2.1) and the sample period T. This study is meant to provide guidance for the selection of these parameters in order to achieve good algorithm performance in terms of optimality and computational time.

2.9.1 Unicycle Steering Model

The unicycle canonical model is a fundamental tool in mobile robotics and control systems to represent and control simple wheeled robots [13]. At its core, the unicycle model is an abstraction that simplifies the motion modeling of a robot or vehicle. The name "unicycle model" is derived from its kinematics that closely resemble that of a unicycle. The model has two primary motion components: translation (moving forward or backward) and rotation (turning left or right).

In a two-dimensional plane, the state of the unicycle can be represented by three variables (x, y, θ), where (x, y) represents the Cartesian position of the robot and θ is the robot's orientation (or heading) with respect to the x-axis. Given v as the linear velocity and ω as the angular velocity, the kinematics of the unicycle model can be defined by

$$\dot{x} = v\cos(\theta), \qquad (2.8)$$
$$\dot{y} = v\sin(\theta), \qquad (2.9)$$
$$\dot{\theta} = \omega. \qquad (2.10)$$

For a sample period T we discretize the model using Euler's method, leading to the equations

$$x(kT + T) = x(kT) + v(kT)\cos(\theta(kT))T, \qquad (2.11)$$
$$y(kT + T) = y(kT) + v(kT)\sin(\theta(kT))T, \qquad (2.12)$$
$$\theta(kT + T) = \theta(kT) + \omega(kT)T. \qquad (2.13)$$

2.9.2 Simulation Preliminaries

SBMPO seeks to find an optimal path from a start state to an end state. When an implicit grid with grid resolution r_g is used as described in Section 2.2.1, suboptimality is inherently introduced. A finer grid resolution enables closer approximation to an optimal path but increases the number of nodes and iterations, prolonging the required time to find a trajectory. On the other hand, a coarser grid can decrease the computational time at some sacrifice of optimality. However, there is also a complex interplay between the choosing of the sample period T and the grid resolution r_g that also contributes to algorithm performance in both optimality and computational time.

To illustrate the interrelationship between the choice of T and r_g, consider translational motion of the unicycle in a SBMPO planning algorithm. Let us assume that the x-grid dimension and y-grid dimension are both equal to ℓ. Also, assume that the velocity $v(kT)$ in (2.11) and (2.12) is sampled and the maximum velocity in the sample is V_{max} and that $\theta(kT)$ is sampled such that the vehicle only moves horizontally, diagonally, or vertically. It follows that when considering child nodes generated using V_{max}, one of the three scenarios in Fig. 2.8 will occur during algorithm execution. As the minimum distance to a new grid cell is $\ell/2$ and the maximum distance is $\ell/\sqrt{2}$ and the distance traveled from the parent node is $V_{max}T$, it follows that Scenario 1 occurs when

$$\ell < \sqrt{2}V_{max}T, \tag{2.14}$$

Scenario 2 occurs when

$$\sqrt{2}V_{max}T \le \ell < 2V_{max}T, \tag{2.15}$$

and Scenario 3 occurs when

$$\ell \ge 2V_{max}T. \tag{2.16}$$

If Scenario 3 occurs, the algorithm ends prematurely, while if Scenario 2 occurs the algorithm may become computationally inefficient.

2.9.3 Description of Simulations

As described above in Section 2.9.2, it is assumed that the x-grid dimension and y-grid dimension are both equal to ℓ and this parameter along with T was varied systematically to access their impact on planner optimality and computation time. The parameter ranges and resolution are given in Table 2.3.

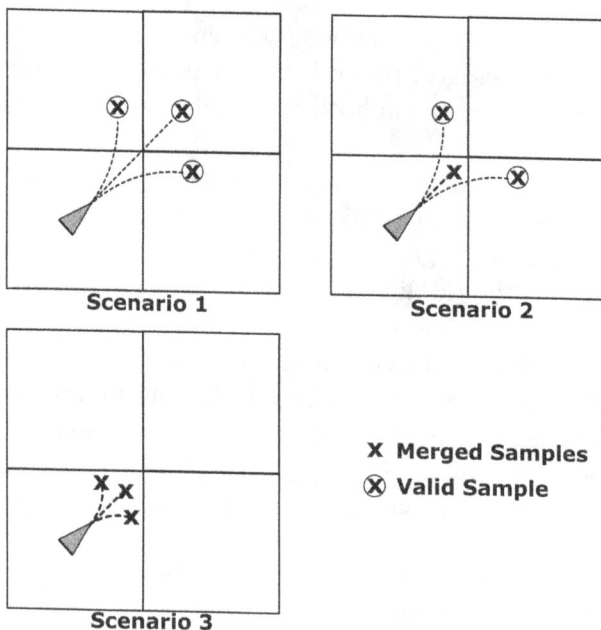

Figure 2.8 These 3 scenarios show child nodes (denoted by **x**) generated from a parent node (denoted by the gray vehicle triangle) by using a maximum velocity sample in cases when the vehicle can only move horizontally, vertically, or diagonally. In Scenario 1, each child node appears in a grid cell that is not that of the parent node and is labeled "valid." In Scenario 2, some of the child nodes appear outside the parent grid cell and others do not and are labeled "merged" since they are indistinguishable from the parent node. In Scenario 3, each of the child nodes appears in the parent grid cell.

The parameters for this benchmark study are given in Table 2.4, where it is seen that the Goal State S_{goal} is solely in terms of the x and y coordinates. Hence, the vehicle can reach the goal irrespective of its angular orientation. In the simulations a single large circular obstacle of radius R_{obs} (see

Table 2.3 Benchmark Parameters

Symbol	Name	Lower Value	Upper Value	Resolution
T	Horizon Time (s)	0.05	0.5	50
ℓ	Grid Resolution (X/Y) (m)	0.1	0.375	50

Table 2.4 and Fig. 2.9) was introduced at the map's center to add planning complexity.

2.9.4 Simulation Results

The contour plots of Figs. 2.10 through 2.12 present simulation results. Referring to Fig. 2.10, Scenarios 1, 2, and 3 of Fig. 2.8 correspond respectively to the region to the right of the black dotted line, the region between the black and gray dotted lines, and the region to the left of the gray dotted line. Fig. 2.11 indicates that, as expected, the number of SBMPO iterations required to find a solution decreases with coarser grid resolution and longer horizon times. However, the constant iteration regions have relatively complex shapes. Fig. 2.12 shows that there is a relatively large region in which the cost is nearly optimal. However, the constant cost regions also have relatively complex shapes.

Overall, the simulation results show that the performance and reliability of the SBMPO planner is dependent on the grid resolution r_g and the sample time T. To achieve the best results in terms of minimizing iterations (and therefore computation time), coarser grid resolutions and larger horizon times are recommended. If achieving an optimal solution is a high priority, then a fine grid resolution with a larger horizon

Table 2.4 Benchmark Constant Parameters

Symbol	Name	Value
N_{iter}	Max Iterations	10^6
N_{gen}	Max Generations	10^3
S_{start}	Start State	$(x : -5\text{m}, y : -5\text{m}, \theta : 0)$
S_{goal}	Goal State	$(x : 5\text{m}, y : 5\text{m}, \theta : *)$
BF	Branchout Factor	10
V_{samples}	Velocity Samples	$[0.5, 1.0]\frac{\text{m}}{\text{s}}$
ω_{samples}	Angular Velocity Samples	$[\pm\frac{\pi}{8}, \pm\frac{\pi}{16}, 0]\frac{\text{rad}}{\text{s}}$
N_{steps}	Integration Steps per Sample	10
R_{body}	Body Radius	0.125m
R_{goal}	Goal Radius	0.5m
M_{bounds}	Map Bounds	$(-10 \leq x \leq 10, -10 \leq y \leq 10)\text{m}$
N_{runs}	Runs per Parameter Set	50
R_{obs}	Obstacle Radius	2.5m
Θ	Grid Resolution (θ)	$\frac{\pi}{16} \times T$

time is recommended. However, these relationships are complex, even for the simple unicycle model. To determine r_g and T for a particular application requires trial and error based on well-chosen simulations. The authors and their co-workers

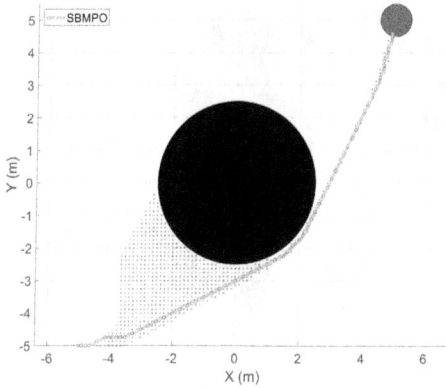

Figure 2.9 Example run of the benchmark with $\ell = 0.15$m and $T = 0.25$s. The highlighted region corresponds to the nodes that appear in the priority queue during the algorithm run.

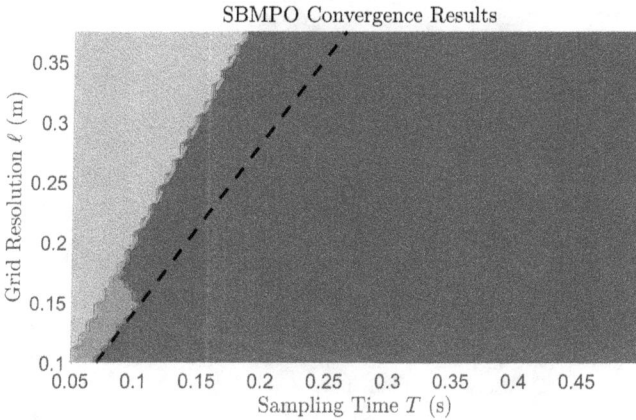

Figure 2.10 Convergence results of the SBMPO planner for varying grid resolutions and sample periods. The dark black region denotes regions where a solution was found, the light black denotes where the iteration limit was reached, and the gray denotes the situation in which all child nodes from S_{start} appeared in the same grid cell as S_{start} (see Scenario 3 in Fig. 2.8), causing algorithm termination.

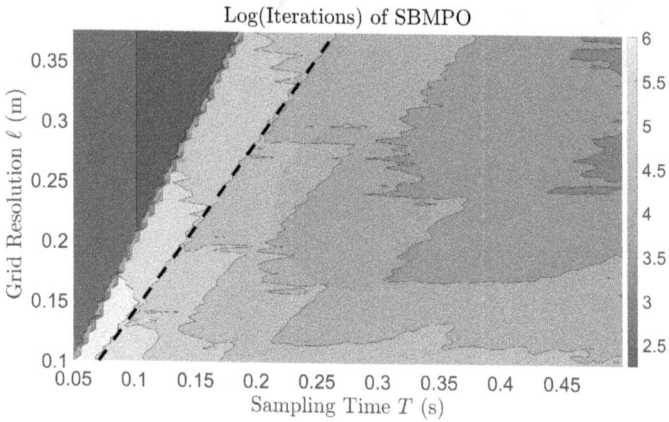

Figure 2.11 Logarithmic scale iterations of the SBMPO planner for varying grid resolutions and sample periods.

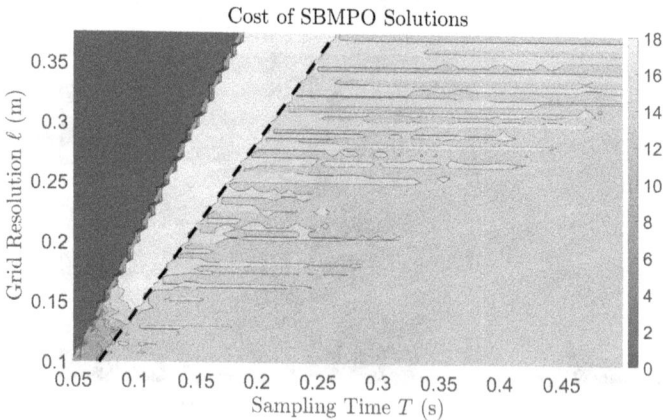

Figure 2.12 Costs of the SBMPO solutions for varying grid resolutions and sample periods. The true optimal cost of this experiment was determined to be approximately 15.2m. The region where the cost is zero represents where the planner could not propagate as it corresponds to the situation in which all child nodes from S_{start} appeared in the same grid cell as S_{start} (see Scenario 3 in Fig. 2.8), causing algorithm termination.

have performed this determination for a large number of problems, including those reported in this book.

2.10 BASELINE COMPARISON OF SBMPO AGAINST RRT*

In this Section, we provide a comparison of SBMPO against the well-established RRT* method, while optimizing a distance cost function.

The model used consists of a holonomic, velocity-driven circular robot of radius $r = 0.2$m. The robot control inputs are the horizontal velocity, $v_x = \{-1.0, 0.0, 1.0\}$m/s, and the vertical velocity, $v_y = \{-1.0, 0.0, 1.0\}$m/s. The starting robot location was selected as $(x_s, y_s) = (-3, -3)$m and the goal as $(x_g, y_g) = (3, 3)$m. The sampling time was chosen as $T =$

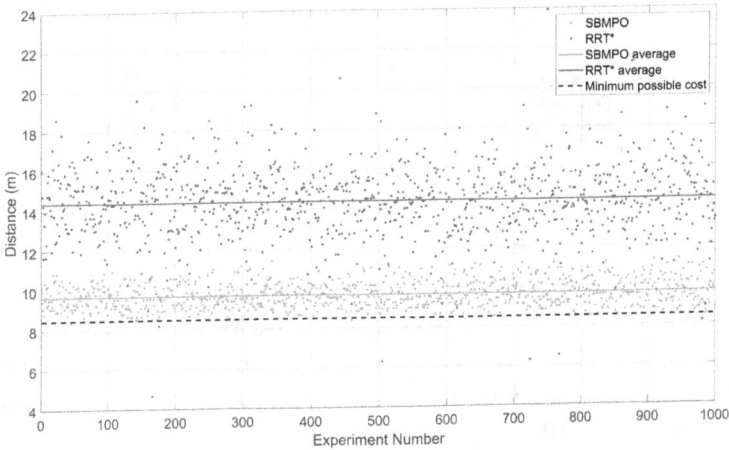

Figure 2.13 Optimality comparison of SBMPO against RRT*. RRT* was allowed to run for the same duration as SBMPO on 1000 randomized obstacle fields. The experiments with cost below the minimum possible distance, correspond to situations for which RRT* did not produce a feasible solution within the planning time.

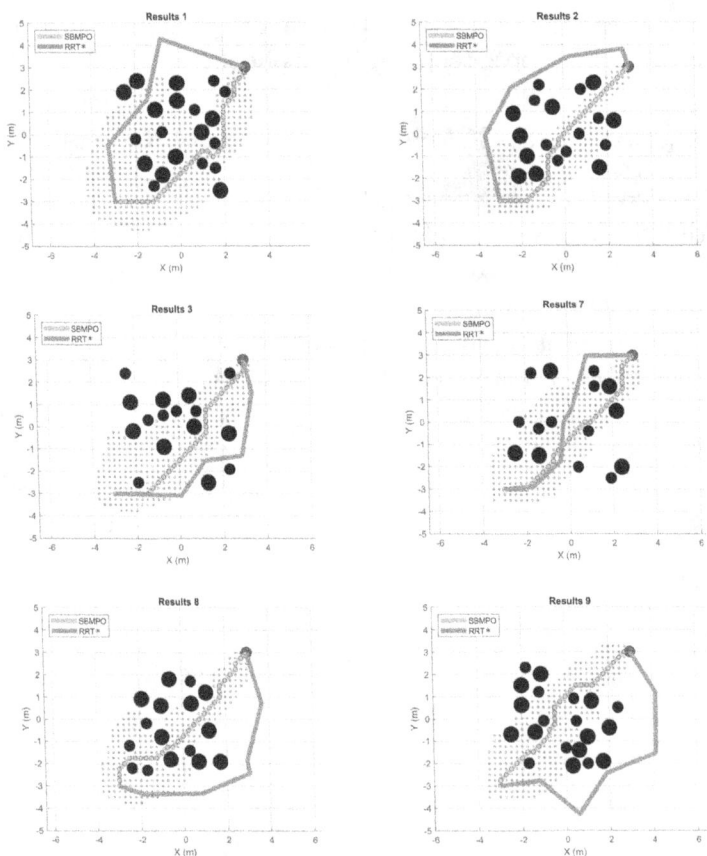

Figure 2.14 Typical results comparing SBMPO and RRT* on randomized scenarios. The result number corresponds to the experiment number from Fig. 2.13, which compares distance optimality for 1000 experiments.

0.25s, and the grid resolution was 0.125m. The obstacle field was constrained to a 10m × 10m square.

A total of 1000 simulations were performed with randomized obstacles as follows: the number of obstacles was randomly selected between 15 and 25, and each obstacle location was chosen randomly inside the bounds of the obstacle map.

Table 2.5 Path Distances Obtained by RRT* and SBMPO in the Scenarios of Fig. 2.14

Result Number	Distance (m)	
	RRT*	SBMPO
1	12.09	9.91
2	11.8	9.22
3	11.07	9.11
7	10.15	9.22
8	11.91	9.36
9	13.66	8.97

Finally, the obstacle size was randomized with radii varying from 0.25m to 0.5m.

To compare the methods, SBMPO was first used to generate a path in each scenario and its computational time was recorded. Then, RRT* was allowed to run for the same amount of time. The results were compared for optimality of the solution (i.e., the total distance of the solution path). Fig. 2.13 shows results comparing the algorithm performance.

Both SBMPO and RRT* were implemented in C++. For RRT* the OMPL [14] motion library was employed as it offers a well-known and optimized implementation of RRT*. The default OMPL parameters of RRT* were used. Fig. 2.14 shows six typical simulation results.

Bibliography

[1] B. Reese, E. Collins, Jr., and F. Alvi, "A nonlinear adaptive method for microjet-based flow separation control,," *AIAA Journal*, vol. 54, pp. 3002–3014, 2016.

[2] B. Reese and E. Collins, Jr., "A graph search and neural network approach to adaptive nonlinear model predictive control,," *Engineering Applications of Artificial Intelligence*, vol. 55, pp. 250–268, 2016.

[3] J. H. Halton, "On the efficiency of certain quasi-random sequences of points in evaluating multi-dimensional integrals," *Numerishe Mathematik*, 1960.

[4] H. Niederreiter, *Random Number Generation and Quasi-Monte Carlo Methods*. SIAM, 1992.

[5] S. Tezuka, *Uniform Random Numbers: Theory and Practice*. Boston, USA: Kluwer Academic, 1995.

[6] C. Ericson, *Real–Time Collision Detection*. Elsevier, 2005.

[7] S. M. LaValle, *Planning Algorithms*. Cambridge University Press, 2006.

[8] R. Dechter and J. Pearl, "Generalized best-first search strategies and the optimality of A*," *Journal of the ACM (JACM)*, vol. 32, no. 3, pp. 505–536, 1985.

[9] J. M. Maciejowski, *Predictive Control with Constraints*. Pearson Education, 2002.

[10] S. Koenig, M. Likhachev, and D. Furcy, "Lifelong planning A*," *Elsevier Science*, May 24 2005.

[11] S. Aine, S. Swaminathan, V. Narayanan, V. Hwang, and M. Likhachev, "Multi-heuristic A*," *The International Journal of Robotics Research*, vol. 35, no. 1-3, pp. 224–243, 2016.

[12] T. Mann, "Application of sampling-based model predictive control to motion planning for robotic manipulators," Master's thesis, Florida State University, Tallahassee, FL, September 2011.

[13] K. Lynch and F. Park, *Modern Robotics Mechanics, Planning, and Control*. Cambridge University Press, 2017.

[14] I. A. Şucan, M. Moll, and L. E. Kavraki, "The Open Motion Planning Library," *IEEE Robotics & Automation Magazine*, vol. 19, pp. 72–82, December 2012.

CHAPTER 3

Development of Heuristics for Direct Generation of Trajectories

Because SBMPO propagates a dynamic (or extended kinematic model, described in Section 4.1.3) to generate a tree, time-dependence of the path is inherent. Hence, SBMPO can directly generate trajectories (as opposed to paths). However, for SBMPO to work efficiently, heuristics that work with dynamic models must be developed – this is not a trivial task. To develop relevant heuristics, we consider the two simple systems shown in Figs. 3.1 and 3.2. They are representative of components of systems that occur respectively in mobile robots and manipulators.

The equation of motion corresponding to Fig. 3.1 is

$$J\ddot{\theta} = \tau, \ \underline{\tau} < \tau < \overline{\tau}. \tag{3.1}$$

Since $x = r\theta$, it follows that

$$\ddot{x} = a, \ \underline{a} < a < \overline{a}, \tag{3.2}$$

DOI: 10.1201/9781003623830-3

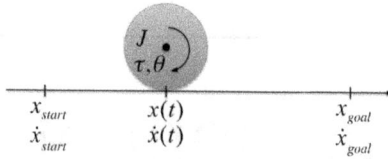

Figure 3.1 Rolling wheel. The inertia J is about the center of mass of the wheel, r is the wheel radius, and the torque τ is bounded by $\underline{\tau} < \tau < \overline{\tau}$.

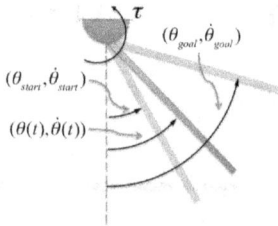

Figure 3.2 Rotating one link manipulator. The inertia of the link about the point of rotation is J and the torque τ is bounded by $\underline{\tau} < \tau < \overline{\tau}$.

where $a = (r/J)\tau$. The equation of motion corresponding to Fig. 3.2 is

$$\ddot{\theta} = \alpha, \ \underline{\alpha} < \alpha < \overline{\alpha}, \tag{3.3}$$

where $\alpha = (1/J)\tau$. which is identical in form to (3.2). Hence, in what follows, we will use (3.2) as a representation of the systems of Figs. 3.1 and 3.2.

It should be noted than in more complex problems, the upper and lower bounds on acceleration may be time varying. In particular,

$$a_{min}(t) \leq \underline{a} \leq a \leq \overline{a} \leq a_{max}(t). \tag{3.4}$$

Hence, the use of the constant bounds $\underline{a}$ and $\overline{a}$ introduces conservatism into the problem, but makes the computation of heuristics computationally tractable.

In practice, the system (3.2) is computer-controlled. Hence, in what follows, the evaluation model is given by the time-discretized double integrator model of Fig. 3.3. In this model,

$$\dot{x}(kT+T) = \dot{x}(kT) + \alpha(kT)T, \tag{3.5}$$

$$x(kT+T) = x(kT) + \dot{x}(kT)T + \frac{1}{2}\alpha(kT)T^2 \tag{3.6}$$

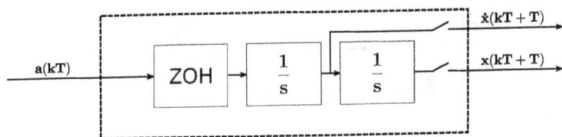

Figure 3.3 Time-discretized double integrator model. This model, which uses a standard zero-order hold at the input and samples the outputs within period T, is used in the simulation results of this section.

3.1 HEURISTIC FOR TIME OPTIMALITY

Time optimal control seeks to optimize the time cost function of Table 2.2,

$$J = \sum_{i=0}^{N-1} t_{k+i}. \tag{3.7}$$

It is assumed that the system is given by (3.2) with $\dot{x}_{goal} = 0$. This optimization problem can be solved using the Maximum Principle [1]. Using a slight generalization of results from [1], the optimal time t_f is computed by solving,

$$t_f^2 + \frac{2q_{2,0}}{u_{min}}t_f + \frac{q_{2,0}^2 + 2(u_{max}-u_{min})q_{1,0}}{u_{min}u_{max}}, \quad \text{if } q_{1,0}+\frac{q_{2,0}|q_{2,0}|}{2u_{max}} > 0, \tag{3.8}$$

$$t_f^2 + \frac{2q_{2,0}}{u_{max}}t_f + \frac{q_{2,0}^2 - 2(u_{max}-u_{min})q_{1,0}}{u_{min}u_{max}}, \quad \text{if } q_{1,0}-\frac{q_{2,0}|q_{2,0}|}{2u_{min}} < 0, \tag{3.9}$$

where u_{min} and u_{max} are the acceleration bounds, and $(q_{1,0}, q_{2,0})$ represent the system's original state. The heuristic $\underline{t}_f$, which was first presented in [2] and also used in [3] and [4], corresponds to the optimal bang-bang control of Fig. 3.4.

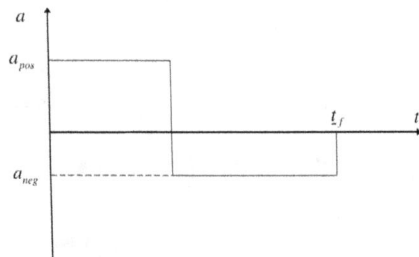

Figure 3.4 Optimal bang-bang Control. The minimum time heuristic corresponds to an optimal bang-bang control strategy.

However, the bang-bang control strategy assumes continuously variable inputs and hence does *not* necessarily equal the control strategy for the time-discretized problem. Also, if the actual system has time-varying bounds as in (3.4), then the optimal solution is generally not bang-bang control.

All results in this section were generated on a computer with an Intel i7-7700k processor running at 4.2GHz using 32GB of RAM. The system operating system ran the default configuration of Ubuntu 18.04 LTS. SBMPO/A* results for minimum time planning are shown in Fig. 3.5 while the parameters used to generate these results are shown in Table 3.1. The solution was computed in 0.062s and the acceleration profile approximates a bang-bang solution. Fig. 3.6 illustrates the SBMPO node expansions with the double integrator models and the minimum time heuristic.

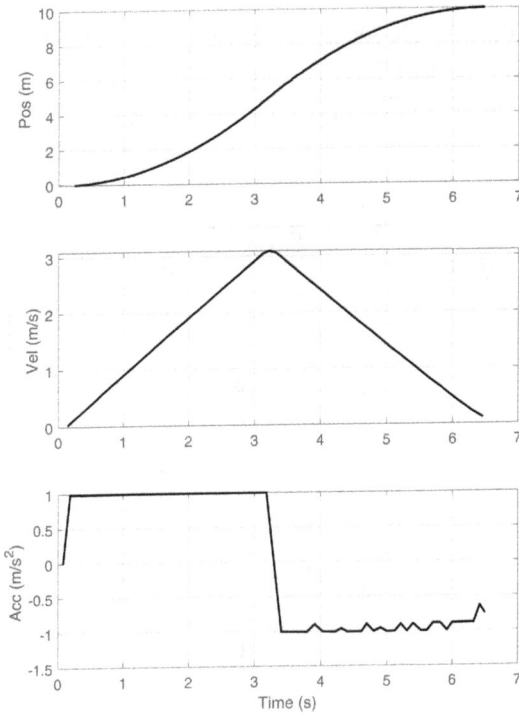

Figure 3.5 SBMPO minimum time solution. The computational time was 0.062s.

3.2 NAIVE HEURISTIC FOR DISTANCE OPTIMALITY

Distance optimal control seeks to optimize the distance cost function of Table 2.2,

$$J = \sum_{i=0}^{N-1} d_{k+i}. \qquad (3.10)$$

It is assumed that the system is given by (3.2) with $\dot{x}_{goal} = 0$. Distance optimal path planning is the most common form of path planning. However, the problem considered here is distance optimal *trajectory* planning, where the goal must be reached at zero velocity. In this section, we use the Euclidean distance to the goal as the heuristic as is often used in A* path

Table 3.1 Parameters Used to Generate the SBMPO Results of Sections 3.1, 3.2, and 3.3

Start	Goal
$x_{start} = 0\text{m}, \dot{x}_{start} = 0\text{m/s}$	$x_{goal} = 10\text{m}, \dot{x}_{goal} = 0$
Acceleration Bounds	**Grid Resolution**
$\underline{a} = -1\frac{\text{m}}{\text{s}^2}, \overline{a} = 1\frac{\text{m}}{\text{s}^2}$	$r_{G,x} = 0.01\text{m}, r_{G,\dot{x}} = 0.001\frac{\text{m}}{\text{s}}$
Branchout Factor	**Sampling Time**
$B = 7$	$T = 0.1\text{sec}$

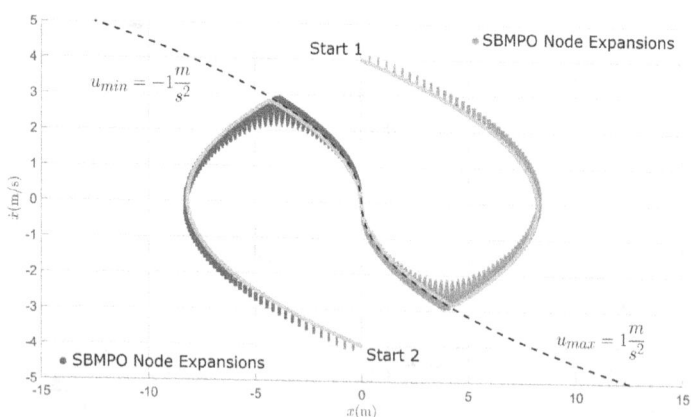

Figure 3.6 Illustration of the guided search performed with the double integrator models and the minimum time heuristic. In this simulation SBMPO used a branchout factor of 7, corresponding to uniformly sampled accelerations in the interval $(-1, 1)$. Two different start states are considered $(x, \dot{x}) = (0, 4)$ and $(x, \dot{x}) = (0, -4)$. In both cases the goal was at $(x, \dot{x}) = (0, 0)$. The remaining SBMPO parameters are the ones of Table 3.1. The gray and black colored points correspond to SBMPO node expansions. The thick black line shows the switching curves of maximum and minimum accelerations.

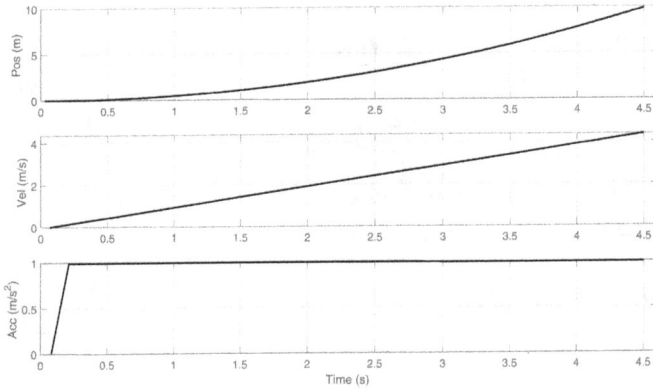

Figure 3.7 SBMPO naive minimum distance solution.

planning. In addition, the heuristic was used to break ties in the priority queue. In particular, if two nodes had equal overall cost, the node with the lower heuristic was placed at the top of the priority queue. This encouraged the planner to produce trajectories with higher velocity.

SBMPO/A* results for planning with the naive heuristic described above are shown in Fig. 3.7 and the parameters used to generate these results are shown in Table 3.1. Note in this case SBMPO does not converge. This is because the heuristic is naive and ignores the intended objective of reaching the goal at zero velocity. Hence, it has no ability to distinguish between approaching the goal at high and low velocities, although to avoid overshooting the goal, lower velocities are needed as the trajectory approaches the goal.

3.3 VELOCITY-AWARE HEURISTIC FOR DISTANCE OPTIMALITY

As in Section 3.2, the objective is to optimize the distance cost function (3.10) and it is assumed the system is given by (3.2) with $\dot{x}_{goal} = 0$. However, to enable efficient trajectory planning, it is desired to develop a heuristic that is *velocity-aware*, i.e., it takes into account both the goal position *and*

Table 3.2 Computation of Velocity-Aware Heuristic. "Same" denotes that the heuristic is not actually dependent upon the relationship between Δx_{min} and $x_{goal} - x(t)$ and has the value given to the left. The following key is used in the table $* = \Delta x_{min} > |x_{goal} - x(t)|$, $** = \Delta x_{min} \leq |x_{goal} - x(t)|$, $A = 2\Delta x_{min} + |x_{goal} - x(t)|$, and $B = 2\Delta x_{min} - |x_{goal} - x(t)|$.

Case	$x(t)$	$\dot{x}(t)$	Δx_{min}	heuristic $*$	heuristic $**$
1	$< x_{goal}$	< 0	$\left\|\frac{-\dot{x}^2}{2\underline{a}}\right\|$	A	Same
2	$< x_{goal}$	≥ 0	$\frac{-\dot{x}^2}{2\underline{a}}$	B	$x_{goal} - x$
3	$> x_{goal}$	≤ 0	$\left\|\frac{-\dot{x}^2}{2\underline{a}}\right\|$	B	$x - x_{goal}$
4	$> x_{goal}$	> 0	$\frac{-\dot{x}^2}{2\underline{a}}$	A	Same

the goal velocity. The heuristic developed in this section was originally developed in [5].

Development of a velocity-aware heuristic makes judicious use of a simple formula from physics. In particular, assume that a particle has acceleration $a = a^*$, where a^* is some constant, and the initial and final velocities are given respectively by $\dot{x}_0$ and $\dot{x}_f$. Then,

$$\dot{x}_f^2 = \dot{x}_0^2 + 2a^*(x_f - x_0). \tag{3.11}$$

If $\dot{x}_f = 0$, then

$$\Delta x = x_f - x_0 = \frac{-\dot{x}_0^2}{2a^*}. \tag{3.12}$$

This equation is used to determine the minimum stopping distance Δx_{min} in Table 3.2. Pseudo-code that implements this heuristic is given in Algorithm 1.

Algorithm 1 Velocity Aware Heuristic for Distance Optimality

1: **if** $x < x_{goal}$ **then**
2:      **if** $\dot{x} < 0$ **then**   ▷ headed away from goal
3:          $\Delta x_{min} = \left| \frac{-\dot{x}^2}{2a} \right|$ ▷ minimum distance to stop
4:          h $= 2\Delta x_{min} + |x_{goal} - x|$
5:      **if** $\dot{x} \geq 0$ **then**   ▷ headed to goal or at rest
6:          $\Delta x_{min} = \frac{-\dot{x}^2}{2a}$ ▷ minimum distance to stop
7:          **if** $\Delta x_{min} > |x_{goal} - x|$ **then**
8:              h $= 2\Delta x_{min} - |x_{goal} - x|$
9:          **else**
10:              h $= x_{goal} - x$
11: **else if** $x > x_{goal}$ **then**
12:      **if** $\dot{x} \leq 0$ **then**   ▷ headed to goal or at rest
13:          $\Delta x_{min} = \left| \frac{-\dot{x}^2}{2a} \right|$ ▷ minimum distance to stop
14:          **if** $\Delta x_{min} > |x - x_{goal}|$ **then**
15:              h $= 2\Delta x_{min} - |x_{goal} - x|$
16:          **else**
17:              h $= x - x_{goal}$
18:      **if** $\dot{x} > 0$ **then**   ▷ headed away from goal
19:          $\Delta x_{min} = \frac{-\dot{x}^2}{2a}$ ▷ minimum distance to stop
20:          h $= 2\Delta x_{min} + |x_{goal} - x|$
     return h

SBMPO/A* results employing the velocity aware heuristic for distance optimality are shown in Fig. 3.8. The parameters used to generate these results are shown in Table 3.1. In this scenario, the planner successfully generated a trajectory to the desired goal configuration in 0.6ms.

3.4 HEURISTIC FOR ENERGY OPTIMALITY

Consider again the system (3.2) with $\dot{x}_{goal} = 0$. The corresponding energy cost function of Table 2.2 is given by

$$J = \sum_{i=0}^{N-1} E_{k+i}, \qquad (3.13)$$

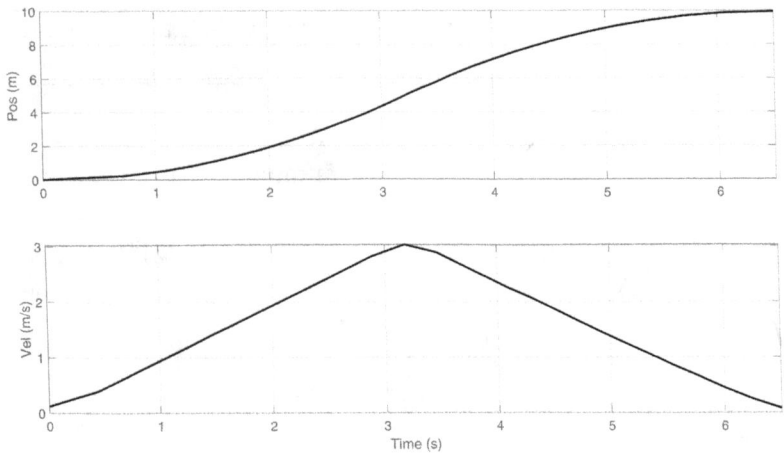

Figure 3.8 SBMPO velocity-aware minimum distance solution.

where

$$E_j = a_j d_j \qquad (3.14)$$

and a_j is the constant force applied from node j to $j + 1$. However, the corresponding optimization problem is ill-posed. To see this note that the system (3.2) is equivalent to applying a force a to a unity mass moving from start to goal on a frictionless surface. Hence, the energy to move the mass to the goal can be infinitesimally small.

Energy optimization makes more sense when friction is in the problem and will be considered in Section 4.3 for a skid-steered vehicle moving in general curvilinear motion at constant velocity.

Bibliography

[1] A. E. Bryson and Y.-C. Ho, *Applied Optimal Control.* Taylor and Francis, 1975.

[2] O. Chuy, J. Emmanuel G. Collins, D. Dunlap, and A. Sharma, "Sampling-based direct trajectory generation using the minimum time cost function," in *13th Inter-*

national *Symposium on Experimental Robotics*, (Quebec, Canada), pp. 651–666, 6 2013.

[3] G. Francis, E. Collins, O. Chuy, and A. Sharma, "Sampling-based trajectory generation for autonomous spacecraft rendezvous and docking," in *AIAA Guidance, Navigation, and Control Conference*, (Boston, MA), 8 2013.

[4] O. Chuy, E. Collins, A. Sharma, and R. Kopinsky, "Using dynamics to consider torque constraints in manipulators planning with heavy loads," *ASME Journal of Dynamic Systems, Measurement, and Control*, 2016.

[5] T. Mann, "Application of sampling-based model predictive control to motion planning for robotic manipulators," Master's thesis, Florida State University, Tallahassee, FL, 2011.

CHAPTER 4

Illustration of SBMPO Motion Planning Using Dynamic Models

The more general scenario for using SBMPO for trajectory planning is described in Fig. 4.1. The problem begins by determining a cost function and the models associated with the problem, which in turn are used to determine the four components on the right hand side of Fig. 4.1. These four components are integral to SBMPO. As shown in Figs. 2.6 and 4.1, direct trajectory planning using SBMPO requires the synergism of various fields: kinematic, dynamic, and power modeling; control systems; and artificial intelligence.

Below, we consider, in a unified fashion, trajectory planning applications of SBMPO appeared in the literature [1, 2, 3, 4, 5, 6]. Each of these applications is described in terms of the cost function and associated models, and then in terms of how these are used to generate the right hand side elements of Fig. 4.1. Finally, the specialization of the 6 key steps of

DOI: 10.1201/9781003623830-4

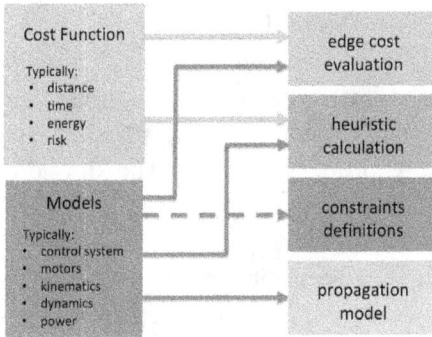

Figure 4.1 SBMPO planning problem formulation.

Section 2.5 to each problem is given. Steps 1 and 2 are always the same regardless of the application, and for computational speed, Step 6 (a) is always used such that the SBMPO algorithm was stopped when the goal is reached. Hence, in all descriptions of the key steps, details are only offered for SBMPO Steps 3–5.

In Section 4.1, SBMPO is used for momentum-based planning for both AGVs and manipulators. In these applications, momentum is needed due to the torque limitations of the robot motors. Propagation of the system dynamic model is avoided as it would require constant propagation of a control system for the wheel or joint torques, which is updated at a much higher rate than the planner, leading to large and impractical computational times. Instead, a double integrator model is used as the propagation model and the dynamic model is used to extract constraints that can be enforced by SBMPO; this extraction is the main focus of this section. In Section 4.2, SBMPO is applied to spacecraft planning in which momentum is important because of the lack of friction when moving through space and the need to control the rendezvous velocity. It is assumed that the torque control system's influence on the dynamic model is small due to the low influence of gravity in that environment. Hence, in this application the dynamic model *is* directly propagated. This particular problem illustrates the performance of SBMPO with a high-dimensional (6-DOF) system with a 12 dimensional state space representation that is used while planning. In Section 4.3 energy-efficient motion planning for skid-steered vehicles is considered. A primary focus is the development of the power model and its integration into planning through the cost function. An extended kinematic model is used as the propagation model. Finally, in Section 4.4, we consider thermally informed motion planning for a legged robot operating in lunar conditions. This system showcases a novel incorporation of a thermomechanical cost function, which is critical for mission endurance and the integration of SBMPO with the Robot Operating System (ROS) and mapping libraries such as Grid Map [7].

4.1 MOMENTUM-BASED PLANNING

Two types of momentum-based planning have been considered in the literature [1, 2]. First, as illustrated in Fig. 4.2, *mobility challenges* are considered here to be segments of terrain that are traversable for a given robot provided it has sufficient momentum when entering the terrain. The underlying assumption is that due to motor torque limitations, the vehicle can only decelerate when traversing the challenging terrain segment. Second, as illustrated in Fig. 4.3, a manipulator with sufficiently high load has a controllability region in which the load can be held statically. To lift the load above the lower region sufficient momentum is required. This is the problem of autonomous manipulator planning with heavy loads.

4.1.1 Dynamic Model

SBMPO enables direct generation of trajectories by propagating a model. This model *can* be a discrete-time representation of a dynamic model of the form,

$$M(q)\ddot{q}(t) + C(q(t), \dot{q}(t)) + G(q(t)) = \tau(t), \qquad (4.1)$$

Figure 4.2 Outdoor mobility challenges. These terrains have the common feature that sufficient momentum may be needed to traverse them due to torque limitations of a mobile robot. (a) Steep Hills (b) Mud, and (c) Thick Vegetation.

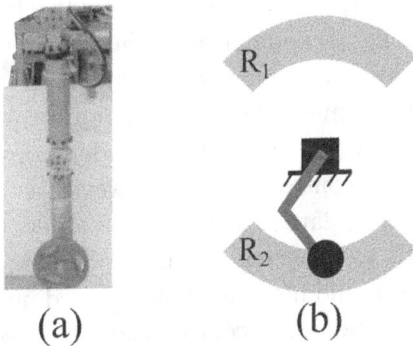

Figure 4.3 (a) 2-DOF Manipulator (b) Controllability Region. For a sufficient load, the 2-DOF manipulator has a limited controllability region.

where q is a vector of generalized coordinates, $M(q)$ is a matrix of masses and inertias, $C(q, \dot{q})$ denotes resistance (typically due to friction), $G(q)$ denotes forces due to gravity, and τ is due to external torques; here, for simplicity it is assumed that only torque motors are used. Although (4.1) does contain the relationship between the torques and vehicle motion, the actual torque vector applied to the robot is determined by the control system.

4.1.2 Control System

A common and effective controller for robots is the computed torque controller [8], which is used in the momentum-based planning described here. The inputs to the controller are the desired acceleration, velocity, and position (i.e., $\ddot{q}_d$, $\dot{q}_d$, and q_d) and the applied torque τ_d is calculated as

$$
\begin{aligned}
\tau_d(t) = M(q)[\ \ddot{q}_d(t) + K_v(\dot{q}_d(t) - \dot{q}(t)) \\
+ K_p(q_d(t) - q(t))\ ] + C(q(t), \dot{q}(t)) + G(q(t)),
\end{aligned}
\tag{4.2}
$$

where $M(q)\ddot{q}_d$ is the feedforward term, $C(q, \dot{q}) + G(q)$ are respectively the friction and gravity compensation terms, $M(q)[K_v(\dot{q}_d - \dot{q}) + K_p(q_d - q)]$ is the feedback term, and $K_v \in R^{n \times n}$ and $K_p \in R^{n \times n}$ are the feedback gains. Typically, the controller is updated at a much faster rate (e.g., 1 kHz) than the desired planner rate (e.g., 10 Hz).

The most direct method of planning for torque limitations is to simply choose the closed-loop model, consisting of the plant (4.1) and the controller (4.2). However, as the controller has a much faster update rate, perhaps 2 orders of magnitude faster, this leads to unacceptably long planning times as the controller model would have to be propagated numerous times for each planner update. A computationally feasible alternative is developed in [1, 2] and relies on a double integrator model.

4.1.3 Double Integrator Model

Consider the acceleration vector $\ddot{q}_d(kT)$. If it is held constant over the interval $[kT, kT + T)$, then from basic physics,

$$\dot{q}_d(kT + T) = \dot{q}_d(kT) + T\ddot{q}_d(kT), \qquad (4.3)$$

$$q_d(kT + T) = q_d(kT) + \dot{q}_d(kT)T + \frac{1}{2}\ddot{q}_d(kT + T)T^2 \qquad (4.4)$$

In matrix form,

$$
\begin{bmatrix} \dot{q}_d(kT + T) \\ q_d(kT + T) \end{bmatrix} = \begin{bmatrix} I & 0 \\ T & I \end{bmatrix} \begin{bmatrix} \dot{q}_d(kT) \\ q_d(kT) \end{bmatrix}
$$
$$
+ \begin{bmatrix} T \\ \frac{1}{2}T^2 \end{bmatrix} \ddot{q}_d(kT), \ k = 1, 2, \cdots,
\qquad (4.5)
$$

Equation (4.5) is called here a *double integrator model* and can be used as the propagation model in SBMPO. It has as its input angular acceleration $\ddot{q}_d(kT)$ and produces the velocity and position terms, $\dot{q}_d(kT + T)$ and $q_d(kT + T)$, also needed by the computed torque controller (4.2).

A traditional kinematic model cannot be used as the propagation model in SBMPO since it does not have time dependence. However, it can be placed in series with the double integration model as shown in Fig. 4.4 to provide a propagation model. Note that the output of the system of this figure is the position vector x_d and velocity vector $\dot{x}_d$ of interest. This series combination is called an *extended kinematic model*. This type of model can be used when the primary interest is the robot kinematics, for example in applications such as parallel parking.

Models can be developed with an arbitrary number of integrators. For example, one integrator would correspond to a velocity input $\dot{q}_d(kT)$ while three integrators would correspond to a jerk input $\dddot{q}_d(kT)$. In Section 4.3 an example of using one integrator is given.

Note that integrator models and extended kinematic models are dynamic models in that the variables q_d varies with time. However, it is not a dynamic model in a more traditional mechanical modeling sense in that it does not contain any information about the influence of forces and torques

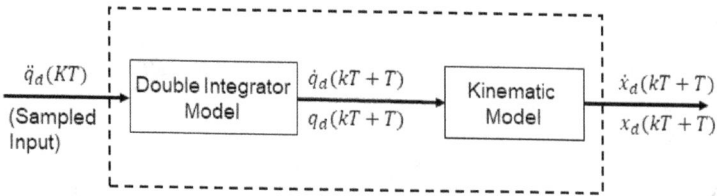

Figure 4.4 Extended kinematic model. This SBMPO propagation model consists of a series combination of the double integration model (4.5) and the appropriate robot kinematics.

on the robotic system. Hence, it could be considered a pseudo-dynamic model. It's value for SBMPO is its simplicity; in particular, it is computationally inexpensive to propagate these models.

4.1.4 Input Constraints

In SBMPO the inputs to the propagation model must be contained in a finite-dimensional set. Typically, this is achieved by providing lower and upper bounds on each element of the input vector. When the input is the acceleration $\ddot{q}_d$, these upper and lower bounds can be chosen as "best-case" constraints. For example, for the momentum-based mobile robot planning problems of [2], the minimum and maximum accelerations are those experimentally observed for the wheels when the robot is moving in a straight line on flat, clear ground under minimum or maximum torque, where the minimum torque is a negative torque (usually the negative of the maximum torque).

Unfortunately, acceleration constraints do not inherently enforce torque constraints, which are the true dynamic constraints. These torque constraints are of the form,

$$-\tau_{max,i}(t) < \tau_i(t) < \tau_{max,i}(t), \ i = 1, 2, \cdots, n. \qquad (4.6)$$

To enforce these constraints, one may ignore the feedback terms in the computed torque controller (4.2) and estimate the torque applied at the beginning of an interval $[kT, kT+T)$

as

$$\tilde{\tau}_d(kT) = M(q)\ddot{q}_d(kT) + C(q_d(kT), \dot{q}_d(kT)) \\ + G(q_d(kT)). \tag{4.7}$$

Then, as discussed in [1], to accommodate the feedback terms in (4.2) and the fact that the torque constraints are only checked at the beginning of the interval $[kT, kT + T)$ instead of along the entire interval, choose $\overline{\tau}_i$ such that $\overline{\tau}_i < \tau_{max,i}$. (The value of $\overline{\tau}_i$ can be determined in part by observing the behavior of the feedback terms as the robot undergoes a variety of maneuvers or by trial-and-error.) Then check whether for $i \in \{1, 2, \cdots, n\}$,

$$\tilde{\tau}_{d,i}(kT) > \overline{\tau}_i \tag{4.8}$$

or

$$\tilde{\tau}_{d,i}(kT) < -\overline{\tau}_i. \tag{4.9}$$

If (4.8) holds, then $\tilde{\tau}_{d,i}(kT) \leftarrow \overline{\tau}_i$ or if (4.9) holds, then $\tilde{\tau}_{d,i}(kT) \leftarrow -\overline{\tau}_i$. Then make the substitution

$$\ddot{q}_d(kT) \leftarrow M(q)^{-1}[\tilde{\tau}_d(kT) - C(q_d(kT), \dot{q}_d(kT)) - G(q_d(kT))]. \tag{4.10}$$

Equations (4.6)–(4.9) are used as part of Step 3 of the Section 2.5 *SBMPO Algorithm*.

4.1.5 Momentum-Based Planning for Mobile Robots Through Mobility Challenges

Traversal of vegetation patches and steep hills can be treated as similar problems. Fig. 4.5 illustrates a wheeled mobile robot approaching these two mobility challenges. The dynamic model (4.1) becomes

$$M\ddot{x} + C(x, \dot{x}) + G(x) = F, \tag{4.11}$$

where M is the vehicle mass, x, $\dot{x}$, and $\ddot{x}$ are the vehicle's linear position, velocity, and acceleration, respectively. F is the tractive force. The frictional term is here given by

$$C(x, \dot{x}) = b(x)\dot{x} + R_r(x), \tag{4.12}$$

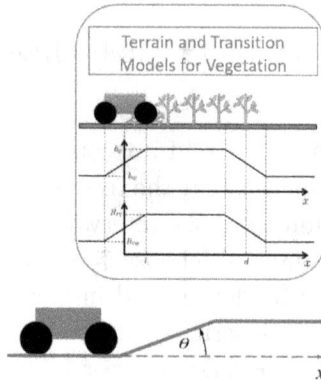

Figure 4.5 Wheeled mobile robot approaching a vegetation patch and a steep hill. Source: [2].

where $b(x)$ is a damping coefficient and $R_r(x)$ represents the rolling resistance. When the vehicle is completely on the hill, the gravitational term is given by

$$G(x) = \frac{W}{2}\sin(\theta), \qquad (4.13)$$

with W representing the robot weight and θ the hill inclination.

A model of the motors is used to compute the maximum torque τ_{max} based on the current wheel angular velocity. Assuming that there is enough friction and a wheel of radius r, the maximum tractive force is

$$F_{max} = 2\frac{\tau_{max}}{r}. \qquad (4.14)$$

Furthermore, since

$$\ddot{x} = \frac{1}{M}[F - C(x, \dot{x}) - G(x)], \qquad (4.15)$$

the maximum acceleration is given by

$$\ddot{x}_{max} = \frac{1}{M}F_{max}. \qquad (4.16)$$

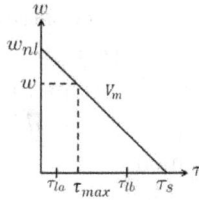

Figure 4.6 Speed vs. torque motor curve used to calculate maximum torque τ_{max}, given the wheel speed ω and assuming a nominal voltage V_m applied to the motor. The motor stall torque is τ_s and its no-load speed ω_{nl}. The values τ_{la} and τ_{lb} represent different torque loads. Source: [2].

Similarly, the minimum acceleration (maximum deceleration) becomes

$$\ddot{x}_{min} = \frac{1}{M}[-F_{max} - C(x, \dot{x}_{max}) - G(x)_{max}], \qquad (4.17)$$

where $\dot{x}_{max}$ is the maximum vehicle speed and yields the greatest resistance $C(x, \dot{x})$ and $G(x)_{max}$ corresponds to the gravitational term when the vehicle is completely on the hill, i.e., $G(x)$ as given by (4.13). Note that the resistance and gravity contribute to vehicle deceleration.

The calculation of τ_{max} is based on a speed vs. torque motor curve shown is in Fig. 4.6. The motor torque depends on the wheel speed ω and the operation voltage of the motor V_m and is given by

$$\tau_{max} = 2\tau_s \left[1 - \frac{\omega}{\omega_{nl}}\right], \qquad (4.18)$$

where ω_{nl} is the no-load speed of the motor and τ_s is the stall torque at the voltage V_m.

The double integrator model in this application is presented in Fig. 4.7. In this system, the input being sampled is the vehicle acceleration $\ddot{x}_d(kT)$, which is then integrated to compute the system outputs (velocity and position) $\dot{x}_d(KT + T)$ and $x_d(kT + T)$.

The fundamental dynamic constraint is that the force $F_d(t)$ required to follow a trajectory denoted by $(\ddot{x}_d(t), \dot{x}_d(t), x_d(t))$ does not exceed the maximum tractive force. In particular, if $(\ddot{x}_d(t), \dot{x}_d(t), x_d(t))$, $t \in [kT, kT + T)$, it is desired to avoid a situation in which

$$F_d = M\ddot{x}_d + C(x_d, \dot{x}_d) + G(x_d) > F_{max}, \qquad (4.19)$$

which would indicate that the trajectory is unfeasible. If (4.19) is true, then to ensure $F_d \leq F_{max}$,

$$\ddot{x}_d \leftarrow \frac{1}{M}[F_{max} - C(x_d, \dot{x}_d) - G(x_d)], \qquad (4.20)$$

and $\dot{x}_d(t)$ and $x_d(t)$ are recomputed as in Fig. 4.7. In practice, determination of $F_d > F_{max}$ is only accomplished at the beginning of the interval $[kT, kT + T)$, i.e., using $(\ddot{x}_d(kT), \dot{x}_d(kT), x_d(kT))$. Note that this process uses the dynamic model without integrating it.

Time optimal control is discussed in Section 3.1. Here, it seeks to optimize the cost function (3.7), where

$$\ddot{x}_d = u; \ \ddot{x}_{min} \leq u \leq \ddot{x}_{max} \qquad (4.21)$$

with initial states $(x_d(t_0), \dot{x}_d(t_0))$ and final states $(x_d(t_f), 0)$. The equations used to compute the corresponding heuristic $\underline{t}_f$ are (3.8) and (3.9) of Section 3.1.

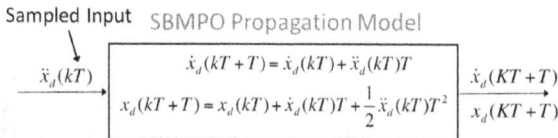

Figure 4.7 Double integrator model used as the SBMPO propagation model for momentum-based traversal of mobility challenges.

Key steps to Momentum-Based Planning for Mobile Robots Through Mobility Challenges

3. *Sample the control space.* Sample the model input $\ddot{x}_d(KT)$. Use (4.19) and (4.20) to respect the maximum tractive force constraints.

4. *Generate neighbor nodes.* Compute $\dot{x}_d(kT + T)$ and $x_d(kT + T)$ using the propagation model shown in Fig. 4.7.

5. *Compute the heuristic for each node and add the node to the graph.* Compute the heuristic $\underline{t}_f$ for the node corresponding to $(x_d(kT + T), \dot{x}_d(kT + T))$ by solving the minimum time problem (4.21). Then, add the new node to the graph.

A typical experimental result is shown in Figs. 4.8 and 4.9. In this experiment the vehicle is tasked with traversing a plastic vegetation patch of known dimensions and frictional properties to reach a desired goal with zero velocity. The robot starts from rest. In this scenario, if the vehicle is simply commanded to go forward at its maximum speed, it becomes

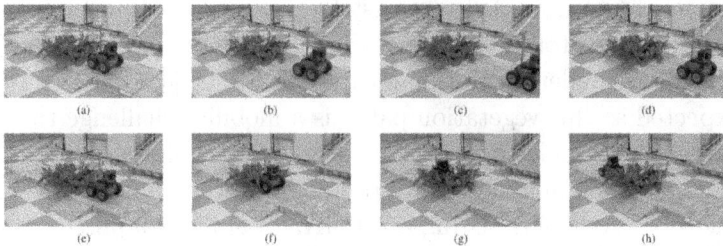

Figure 4.8 Vehicle executing the planned trajectory to overcome the vegetation mobility challenge. The vehicle had to back up (a)–(c) to gather momentum and was able to stop at the goal with zero velocity (h). Without momentum, the vehicle gets immobilized in the vegetation due to its limited torque. Source: [2].

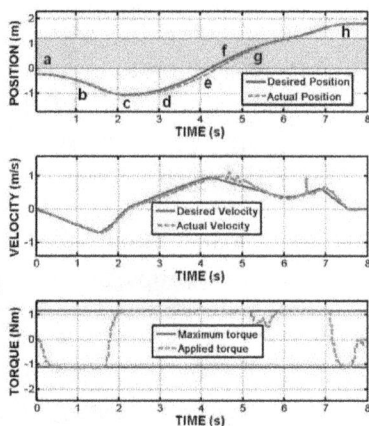

Figure 4.9 Resultant position, velocity, and motor torque trajectory profiles planned by SBMPO. The labels a–h correlate with the snapshots of Fig. 4.8. The black rectangle of the position vs. time graph illustrates the vegetation patch. Source: [2].

immobilized due to its limited torque [2]. The frames from Fig. 4.8 are correlated with the labels a through h shown in the position profile of Fig. 4.9. SBMPO plans a trajectory that forces the vehicle to back up and then accelerate to gather the momentum needed to cross the patch. It is important to note from the torque profile of Fig. 4.9 that the motor torque was in fact under saturation for most of the trajectory. This is expected as this vegetation patch is a mobility challenge that causes the vehicle to decelerate. Fig. 4.9 also shows that the planned position and velocity profiles are properly tracked by the robot controller. Using a 3.4 GHz Intel i7-2600 processor, a branch out factor of $B = 10$, and a sampling frequency of 20Hz, the average computation time for diverse experimental scenarios was 0.251s [2].

4.1.6 Momentum-Based Planning for Manipulators

As explained in Section. 4.1, a manipulator needs to employ its momentum when trying to lift a heavy load to a final configuration that requires the manipulator to operate through non-controllable regions.

Hereafter, we consider an n-DOF manipulator. In addition, referring to the dynamic model (4.1), $(q, \dot{q}, \ddot{q})$ denotes the vector of n joint angles and its derivatives, and τ is the vector of joint torques. Similarly, $(q_d, \dot{q}_d, \ddot{q}_d)$ is the vector of desired joint angles and its derivatives. Since this problem is very similar to the one of Section 4.1.5, here, we simply indicate the key steps followed by SBMPO. For further details refer to [1].

Key Steps to Momentum-Based Planning for Heavy Lifting Trajectory Planning

Referring to Section 2.5, the key steps are as follows.

3. *Sample the control space.* Sample the model input $\ddot{q}_d(KT)$. If for any $i \leq n$, $-\tau_{max,i} < \tau_{d,i}(kT) < \tau_{max,i}$ is violated, then use (4.10) to respect the torque constraints.

4. *Generate neighbor nodes.* Compute $\dot{q}_d(kT + T)$ and $q_d(kT + T)$ using the double integrator model (4.5).

5. *Compute the heuristic for each node and add the node to the graph.* For $i \in \{1, \dots, n\}$ compute the solution $\underline{t}_{f,i}$ to the minimum time problem,

$$\ddot{q}_{d,i} = u_i, \quad -a_i \leq u_i \leq b_i;$$
$$q_{d,i}(0) = q_{d,i}(kT), \dot{q}_{d,i}(0) = \dot{q}_{d,i}(kT), \quad (4.22)$$

where the goal is

$$q_{d,i}(t_f, i) = q_{d,i,goal}, \quad \dot{q}_{d,i}(t_{f,i}) = 0. \quad (4.23)$$

The heuristic $\underline{t}_f$ is then given by

$$\underline{t}_f = max\{\underline{t}_{f,1}, \dots, \underline{t}_{f,n}\}. \quad (4.24)$$

Notice that the maximum value of the set of the n minimum time problems is selected to make $\underline{t}_f$ optimistic. Then, add the new node to the graph.

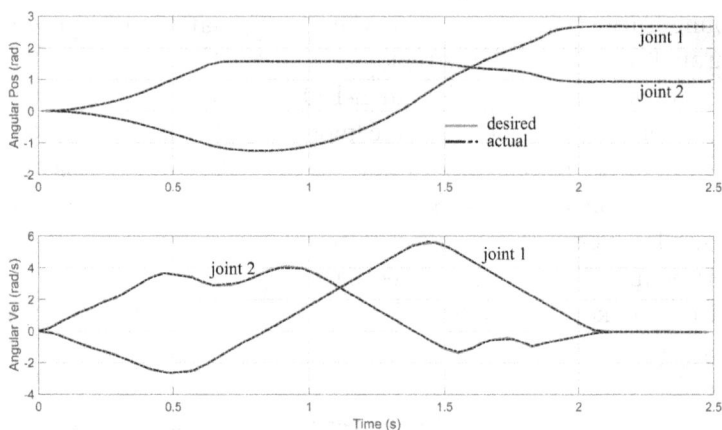

Figure 4.10 Simulation results: Trajectory planning and tracking results for 2.27kg (5lb) load. The desired and actual angular positions and velocities were virtually identical.

Experimental results of momentum-based planning for a 2-DOF manipulator with a load of 5 lb are shown in Figs. 4.10 and 4.11 where $n = 2$, $a_i = b_i = 10$ rad/sec^2, and $\tau_{max,i} = 10$Nm. The manipulator's task was to move from $(0.0, -0.675)$ [m] to a goal position $(0.0, 0.65)$ [m]. Using an iCore 7 Duo 2.5 GHz processor, SBMPO computed a trajectory in 22.8s with a 50Hz update rate. It is important to note that the increased computational time is due to the fact that the manipulator had a very limited control authority to handle this heavy load.

Fig. 4.10 shows the simulation result as the trajectory is fed to the low-level tracking controller. Fig. 4.11 (a) shows a snapshot of the end effector's path from start to goal and it can be noticed that the manipulator moves in the opposite direction to gain the momentum needed to traverse the non-controllable region. Fig. 4.11 (b) shows the experimental result as the trajectory is fed to the tracking controller. Based on the desired and actual joint positions, the average tracking error for joint one is 0.065rad and joint two is 0.028rad.

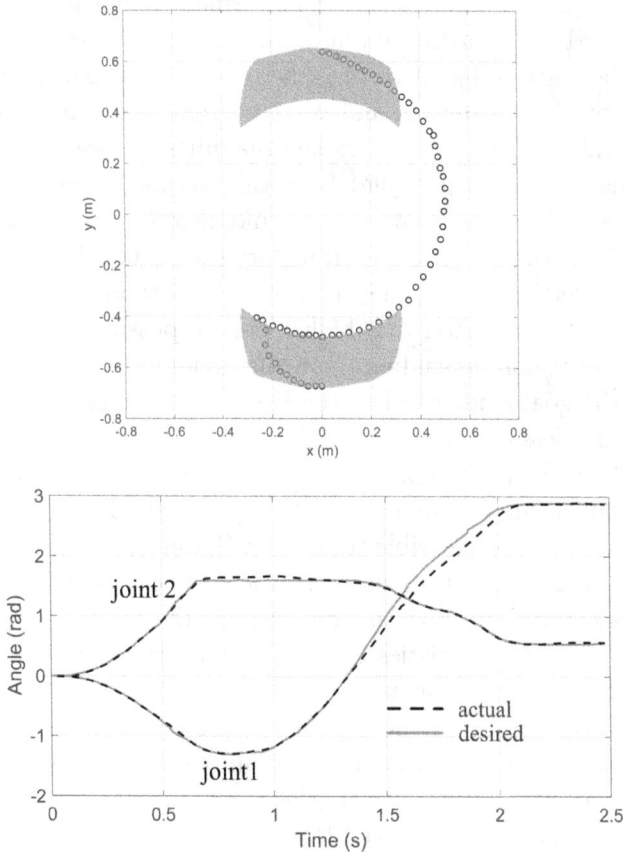

Figure 4.11 Top: End-effector's path of the 2-DOF manipulator with 2.27kg (5lb) load. Bottom: experimental results. Trajectory planning and tracking results for 2.27kg (5lb) load. The desired and actual angular positions closely coincide. The average tracking error for joint one is 0.065rad and joint two is 0.028rad.

To significantly reduce the planning time, learning from baseline trajectories was introduced [1]. The baseline trajectories must be from planning results for a load greater than or equal to the current load. To learn from baseline trajectories, the following modifications were performed. Whereas the standard SBMPO algorithm initializes Q, the priority queue with only the start and goal vertices, here Q is initialized with the estimate of cost-to-goals and vertices from a baseline trajectory. This change accelerates the planning process by using information (when possible) from the baseline trajectory. In practice, this amounts to using the momentum characteristics of the first part of a baseline trajectory (i.e., some of the initial baseline vertices) to enable the manipulator to move from the lower reachable region to a higher region, possibly the upper reachable region depending upon the planning goals.

If the goal is identical to the baseline trajectory goal and the load is less than or equal to that of the baseline trajectory, SBMPO will simply reproduce the baseline trajectory since it has momentum characteristics that enable the load to reach the goal, i.e., it is a feasible trajectory. However, when the goal differs from that of the baseline trajectory, the latter part of the trajectory will deviate from that of the baseline trajectory. If the characteristics of the baseline trajectory are not useful to the current planning task, then SBMPO will give all or most of vertices of the baseline trajectory high costs and essentially plan as if the standard initialization were used. Table 4.1 shows that by using learning, the planning times can be significantly improved [1].

4.2 MINIMUM TIME PLANNING FOR AUTONOMOUS SPACECRAFT

An important problem in aerospace consists in having a spacecraft autonomously planning a trajectory to a non-cooperating piece of debris in the midst of a debris field. This rendezvous maneuver should determine the trajectory and control inputs necessary to achieve motion from an initial to a final configuration, should be made as quickly as possible (i.e., a minimum time maneuver), and should end in zero relative velocity.

Table 4.1 Acceleration of Planning Times via Learning. Baseline trajectory was for a 10lb load and the experiments were run for a 5lb load.

Goal	Without Learning (s)	With Learning (s)	Nodes Reused
(0.0, 0.65)	22.772	0.018	110
(0.10, 0.60)	18.683	0.017	89
(0.30, 0.60)	34.771	0.009	89
(0.30, 0.50)	14.723	0.019	79
(0.25, 0.50)	9.877	0.011	79
(-0.1, 0.60)	11.223	8.180	108
(-0.2, 0.60)	16.567	8.550	108
(-0.3, 0.60)	43.789	8.994	109
(-0.3, 0.50)	11.796	9.775	109
(-0.4, 0.50)	38.581	9.787	109

4.2.1 Spacecraft Dynamics

The continuous, 6 degrees of freedom time dynamics that describes the relative motion of the spacecraft with respect to the target can be cast in control affine form as

$$
\begin{bmatrix} \dot{v} \\ \dot{r} \\ \dot{\omega} \\ \dot{q} \end{bmatrix} = \begin{bmatrix} \tilde{0} \\ \dot{v} \\ -J^{-1}\omega \times J\omega \\ \frac{1}{2}\Omega(\omega)q \end{bmatrix} + \begin{bmatrix} \frac{1}{m}\Theta^T(q)u(t) \\ \tilde{0} \\ J^{-1}\tau(t) \\ \tilde{0} \end{bmatrix}, \quad (4.25)
$$

where r and v are the position and velocity vectors of the spacecraft with respect to the target, $\omega = [\dot{\phi}\ \dot{\theta}\ \dot{\psi}]$ is the angular velocity vector in the spacecraft body frame, $q = [q_0\ q_1\ q_2\ q_3]$

is the rotation quaternion vector, Ω is the kinematic quaternion matrix, m the mass, J the inertia matrix, $u(t)$ the thrust control input vector, $\tau(t)$ the torque control input vector, $\tilde{0}$ is the zero vector, and Θ the rotation quaternion vector given by

$$
\Theta(q) = \begin{bmatrix} (1 - 2q_2^2 - 2q_3^2) & 2(q_1q_3 + q_0q_3) & 2(q_1q_3 - q_0q_2) \\ 2(q_1q_2 - q_0q_3) & (1 - 2q_1^2 - 2q_3^2) & 2(q_2q_3 + q_0q_1) \\ 2(q_1q_3 + q_0q_2) & 2(q_2q_3 - q_0q_1) & (1 - 2q_1^2 - 2q_3^2) \end{bmatrix}.
$$
(4.26)

Using the vehicle dynamics and the process detailed in [4, 5], bounds on vehicle linear acceleration are

$$
-\frac{1}{m}u_{i,max} \le a_i \le \frac{1}{m}u_{i,max}, \ i = 1, 2, 3.
$$
(4.27)

Similarly, the bounds on angular acceleration are

$$
-\frac{1}{\sigma}\tau_{i,max} \le \dot{\omega}_i \le \frac{1}{\sigma}\tau_{i,max}, \ i = 1, 2, 3,
$$
(4.28)

where the spacecraft is assumed symmetric with inertia matrix $J = \sigma I$, and $\sigma > 0$.

For the propagation model used by SBMPO, the discretized spacecraft dynamics are required and are given by

$$
\begin{bmatrix} v(kT + T) \\ r(kT + T) \\ \omega(kT + T) \\ q(kT + T) \end{bmatrix} = \begin{bmatrix} v(kT) \\ r(kT) \\ \omega(kT) \\ q(kT) \end{bmatrix} + \begin{bmatrix} \tilde{0} \\ Tv(kT) \\ -TJ^{-1}\omega(kT) \times J\omega(kT) \\ \frac{1}{2}T\Omega(\omega)q(kT) \end{bmatrix}
$$

$$
+ \begin{bmatrix} \frac{1}{m}T\Theta^T(q(kT))u(kT) \\ \tilde{0} \\ TJ^{-1}\tau(kT) \\ \tilde{0} \end{bmatrix}.
$$
(4.29)

Key Steps in SBMPO for Autonomous Spacecraft Rendezvous

Referring to Section 2.5, the key steps are as follows.

3. *Sample the control space.* Sample the thrusts $u(kT)$ and the torques $\tau(kT)$.

4. *Generate neighbor nodes.* Compute the translational position $r(kT + T)$ and the rotation matrix $\Theta(q(kT + T))$ using the propagation model (4.29)

5. *Compute the heuristic for each node and add the node to the graph.* Compute the heuristic t_f for the node corresponding to $(r(kT + T),\ v(kT + T),\ \Theta(q(kT + T)),\ \omega(kT + T))$ by first solving for $i = 1, 2, 3$ the minimum time problems,

$$\ddot{r}_i = \frac{1}{m}u_i, -u_{i,max} \leq u_i \leq u_{i,max};$$
$$r_i(0) = r_i(kT);\ \dot{r}_i(0) = v_i(kT); \qquad (4.30)$$
$$r_i(t_{f,i}) = r_{i,goal},\ v_i(t_{f,i}) = 0.$$

Then, solving the 3 minimum time problems,

$$\ddot{\phi} = -\frac{1}{\sigma}\tau_1,\ -\tau_{1,max} \leq \tau_1 \leq \frac{1}{\sigma}\tau_{1,max};$$
$$\phi(0) = \phi(kT);\ \dot{\phi}(0) = \omega_1(kT), \qquad (4.31)$$
$$\phi(t_{f,4}) = \phi_{goal}, \dot{\phi}(t_{f,4}) = 0,$$

$$\ddot{\theta} = -\frac{1}{\sigma}\tau_2,\ -\tau_{2,max} \leq \tau_2 \leq \frac{1}{\sigma}\tau_{2,max};$$
$$\theta(0) = \theta(kT);\ \dot{\theta}(0) = \omega_2(kT), \qquad (4.32)$$
$$\theta(t_{f,5}) = \theta_{goal},\ \dot{\theta}(t_{f,5}) = 0$$

$$\ddot{\psi} = -\frac{1}{\sigma}\tau_3,\ -\tau_{3,max} \leq \tau_3 \leq \frac{1}{\sigma}\tau_{3,max};$$
$$\psi(0) = \psi(kT);\ \dot{\psi}(0) = \omega_3(kT), \qquad (4.33)$$
$$\psi(t_{f,6}) = \psi_{goal}, \dot{\psi}(t_{f,6}) = 0,$$

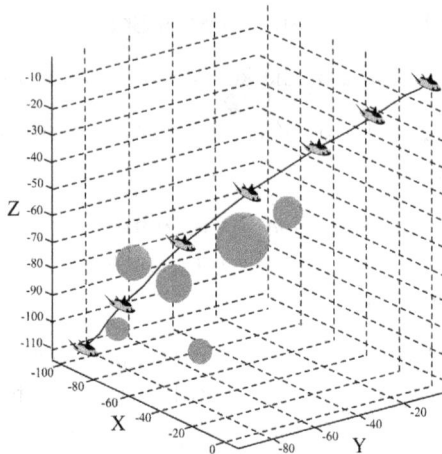

Figure 4.12 SBMPO solution to the rendezvous problem in a cluttered environment.

The heuristic $\underline{t}_f$ is then given by

$$\underline{t}_f = max\{\underline{t}_{f,1}, \cdots, \underline{t}_{f,6}\} \qquad (4.34)$$

Then, add the new node to the graph.

Simulation results of time-optimal, collision-free trajectories of a rendezvous maneuver in a cluttered environment are presented in Figs. 4.12–4.13. In this simulation, seven randomly sized and randomly-positioned spherical objects were included as obstacles impeding the path toward the target.

These simulations were completed on a laptop with a 2.4GHz Intel Core-2 Duo processor and 16GB of DDR3 RAM. In the simulation, the spacecraft is initially disoriented with respect to the target with Euler angles of $(-45.0°, -25.0°, 0.0°)$ and is positioned at $(-101.0m, -87.5m, -111.2m)$. In this configuration, the spacecraft is trailing the target, which is located at the frame of origin. In order to rendezvous with the target, the goal position and orientation should be coincident with the origin frame. As predicted, the resultant trajectory corresponds to

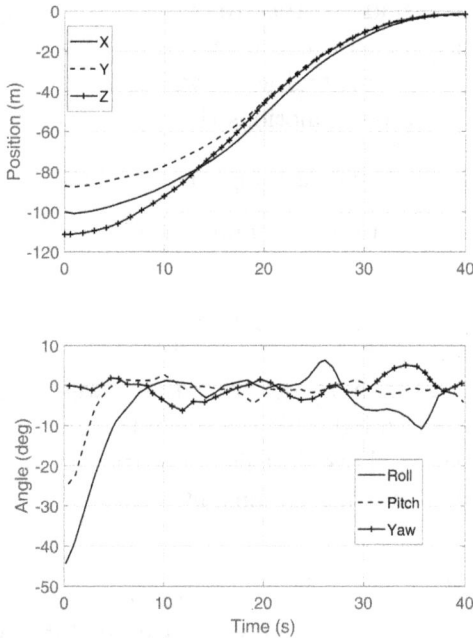

Figure 4.13 Parameter profiles of the SBMPO solution for a minimum time spacecraft rendezvous trajectory in an uncluttered environment. Top: relative position components. Bottom: Euler angle components.

the optimal bang-bang control sequence with some expected divergence due to the presence of impeding obstacles. The collision-free rendezvous trajectory, corresponding to a path length of 179.2m, was generated in 2.23s. For further details, including a discussion about replanning, refer to [5].

4.3 ENERGY EFFICIENT PLANNING FOR SKID-STEERED VEHICLES

Energy efficient planning is particularly important for skid-steered platforms that, due to their mechanical construction, must slip and/or skid in order to turn. The slippage of the wheels or tracks imposes high energetic costs to turning maneuvers. Because of this and as shown in [3], energy efficient trajectories tend to favor motions that minimize turns.

4.3.1 Cost Function and Assumptions

The energy cost function to minimize is given by

$$J = \sum_{i=0}^{N-1} E_{k+i}, \tag{4.35}$$

where $E_j = P_j T$ with P_j denoting the power consumption and T the sample period. Different from the motion planning examples described above, in this application it was assumed that the vehicle is moving at constant speed.

4.3.2 Power Consumption

To estimate energy consumption for a robot maneuver, it is required to develop a model which takes into account both the actuators and the wheel terrain interaction.

Referring to Fig. 4.14, the power for the left (P_l) and right (P_r) wheels can be computed as

$$P_l = \frac{\tau_l \omega_l}{\eta_l} + \left(\frac{\tau_l}{K_t g_l \eta_l} \right)^2 R_e, \tag{4.36}$$

$$P_r = \frac{\tau_r \omega_r}{\eta_r} + \left(\frac{\tau_r}{K_t g_r \eta_r} \right)^2 R_e, \tag{4.37}$$

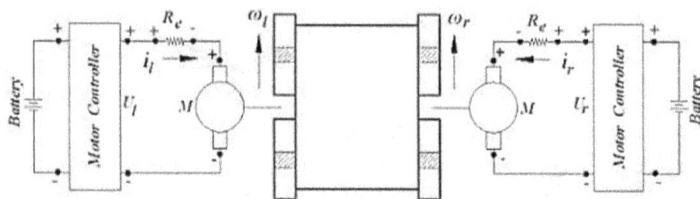

Figure 4.14 Power modeling for skid-steered vehicles. Source: [3].

where τ_* denotes torque, η_* motor efficiency, g_* gear ratio, K_T is the torque constant of the motor, ω_* is the wheel angular velocity, and R_e is the electrical resistance of the motor. The total power is then computed using

$$P(\tau_l, \tau_r, \omega_l, \omega_r) = \sigma(P_r) + \sigma(P_l), \qquad (4.38)$$

where

$$\sigma(Q) = \begin{cases} Q, & Q \geq 0 \\ 0, & Q < 0. \end{cases} \qquad (4.39)$$

As observed in (4.36) and (4.37), the power consumed is dependent on the wheel torques, which in turn are a function of the surface and wheels properties, the robot payload, and the maneuver being performed by the vehicle. Typical torque vs. turn radius curves for the FSU-Bot of Fig. 4.15 moving on a vinyl surface at constant linear velocity are shown in Fig. 4.16.

Figure 4.15 The FSU-BOT skid-steered robot. Source: [3].

An analytical derivation of the torque vs. turn radius curves is detailed in [9]. Real-time adaptation of these curves to different terrain surfaces is presented in [10].

4.3.3 Heuristic Calculation

An optimistic heuristic of the energetic cost to go is computed assuming constant robot speed v and respecting the robot and motor dynamic models. The heuristic is computed as

$$h = P_\infty \Delta t, \qquad (4.40)$$

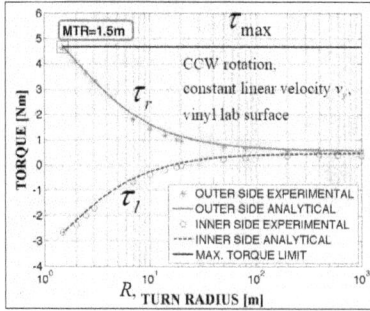

Figure 4.16 Torque vs. turn radii curves for the FSU-BOT on a vinyl surface moving at a constant speed of 0.2m/s. Source: [3].

where P_∞ is the power consumed assuming that the robot is moving straight, i.e., with a turn radius $R = \infty$ toward the goal and at constant velocity v_y. Under these conditions, the wheel angular velocities are $\omega_v = \omega_l = \omega_r$. Therefore, using (4.38), the power consumption becomes $P_\infty = P(\tau_l = \tau_\infty, \tau_r = \tau_\infty, \omega_l = \omega_v, \omega_r = \omega_v)$. The time $\Delta t = \frac{d_g}{v}$ in (4.40) is the time needed to move from the current robot position to the goal, which is at distance d_g.

4.3.4 SBMPO Cost Calculation

The process to estimate the energetic cost incurred by the vehicle moving at velocity v_y and the sampled angular velocity $\dot{\Psi}(kT)$ is summarized in Figs. 4.17–4.18. Given the sampled angular velocity $\dot{\Psi}_k$, the robot is simulated forward using

$$
\begin{bmatrix} X_{G,k+1} \\ Y_{G,k+1} \\ \theta_{G,k+1} \end{bmatrix} = \begin{bmatrix} X_{G,k} + v_y \cos(\theta_{G,k})T \\ Y_{G,k} + v_y \sin(\theta_{G,k})T \\ \theta_{G,k} + \dot{\Psi}_k T \end{bmatrix} \qquad (4.41)
$$

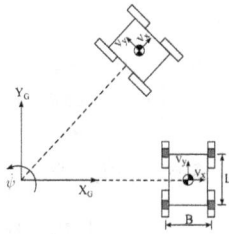

Figure 4.17 Skid-steered vehicle performing a turn with forward velocity v_y and angular velocity $\dot{\Psi}$. The shaded regions represent the contact patch of the wheels. Source: [3].

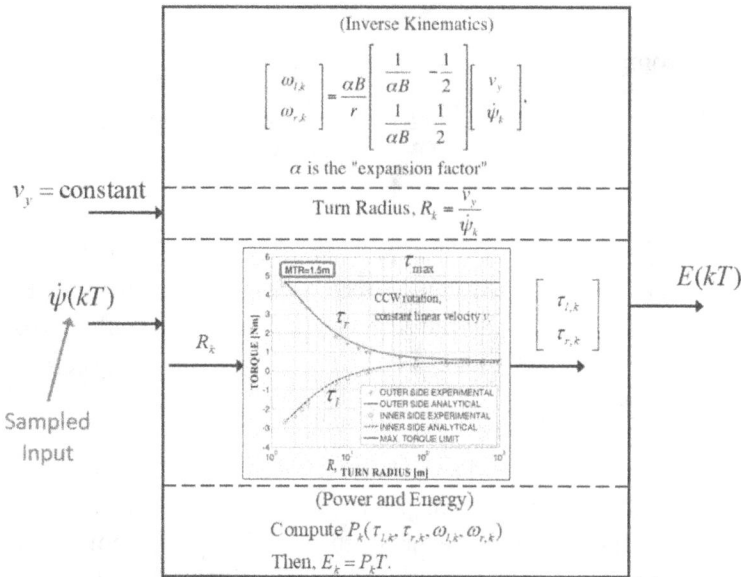

Figure 4.18 Process to estimate energetic cost for a skid-steered moving with constant velocity v_y and angular velocity $\dot{\Psi}(kT)$. Notice that the inverse kinematics involves an expansion factor α which captures the under steering exhibit by these vehicles.

Note that this SBMPO propagation model corresponds to a special case of a single integrator model, where the velocity vector is $[v_y \; \dot{\Psi}_k]^T$ with v_y held constant and only $\dot{\Psi}_k$ sampled.

Key Steps to Energy Efficient Motion Planning

Referring to Section 2.5, the key steps are as follows.

3. *Sample the control space.* Sample the model input $\dot{\Psi}(kT)$ and calculate the turn radius $R(kT) = \frac{v_y}{\dot{\Psi}(kT)}$ and check that it does not violate the minimum turn radius constraint (i.e., the sharpest turn radius that can be commanded without incurring torque saturation). If it does, mark the node as invalid and return to the previous step. During this step the energetic cost $E(kT)$ is also computed.

4. *Generate neighbor nodes.* Using the propagation model (4.41), the new robot position $(X_{G,k+1}, Y_{G,k+1})$ is estimated.

5. *Compute the heuristic for each node and add the node to the graph.* From the updated robot pose, the distance to goal $d_{g,k+1}$ is computed and the heuristic is calculated using $h(k+1) = P_\infty \frac{d_{g,k+1}}{v_y}$. Then, add the new node to the graph.

Fig. 4.19 shows simulation results comparing distance vs. energy optimal motion planning with the FSU-BOT. As shown in the figure, the energy efficient trajectories minimize the sharp turns. Notice that there are significant energetic savings with a relatively small penalty on distance traveled. For detailed experimental results, refer to [3].

4.4 THERMALLY INFORMED MOTION PLANNING

Mission endurance is a critical factor for mobile robots deployed for space exploration and is highly correlated with

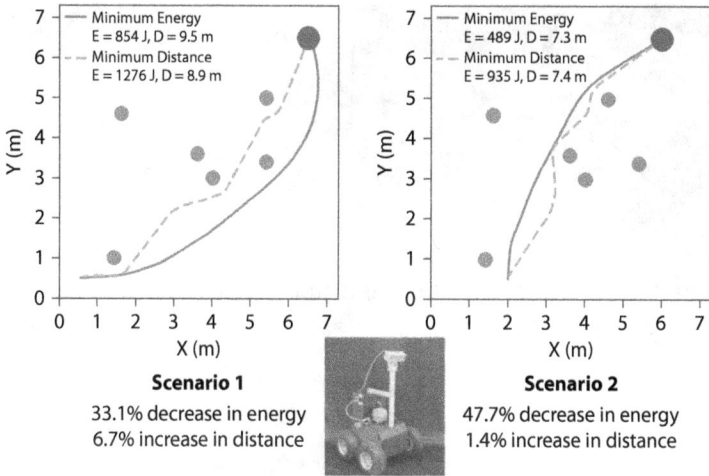

Figure 4.19 Comparison of SBMPO trajectories based on minimizing energy (labeled *Minimum Energy* and minimizing distance (labeled *Minimum Distance*). The computation times were: Scenario 1 (minimum distance: 0.55s, minimum energy: 0.87s); Scenario 2 (minimum distance: 0.56s, minimum energy: 0.18s). Source: [3].

thermoregulation and exposure to solar radiation, which affect both the robot integrity and its access to power. There is an increased interest in developing temperature aware motion planning strategies due to efforts from governments and private companies to expand exploration of the moon and other celestial bodies.

Here, we illustrate how SBMPO is employed in conjunction with thermal maps of celestial bodies to generate thermally aware motion. The work is applied to the recently developed Extreme Terrain Quadruped ETQuad, shown in Figs. 4.20 and 4.21. For further details on the generation of thermal maps and detailed thermomechanical modeling, refer to [6].

Figure 4.20 Extreme terrain quadruped ETQuad. Source: [6].

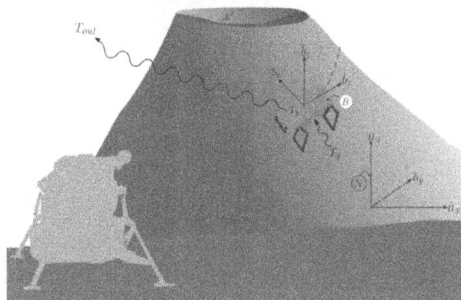

Figure 4.21 ETQuad robot ascending a crater in an energeti-cally efficient manner (dashed black path). There is solar ir-radiance G and the robot body has temperature T_r. There are radiation effects with the ground, at temperature T_g and outer space at temperature T_{out}. There is also heat conduction between the robot and the ground through its 4 legs. The in-ertial frame of reference is denoted by N and the robot frame is B. Source: [6].

4.4.1 Thermal Modeling of ETQuad

ETQuad has mass $m = 6$kg and a geometry approximated by a rectangular parallelepid. The area of the top surface is $A_r = 0.08$m^2. The robot has an estimated emissivity $\epsilon_r = 0.15$. The leg geometry is characterized by a net cross sectional area $A_{clegs} = 5 \times 10^{-4}$m^2 and a leg extension $L_{leg} = 0.2$m. The thermal conductivity of the legs is $k = 170$W/m $\cdot$ K

The main heat transfer mechanisms are radiation between the robot and the ground, radiation between the robot and outer space, and heat conduction between the legs to the ground. A thermal energy balance on the robot is given by

$$
\begin{aligned}
mc_p^r \frac{dT_r}{dt} = & \dot{Q}_{gen}''' V_r + G_N A_r + \epsilon_g \sigma A_r T_g^4 + \epsilon_{out} \sigma A_r T_{out}^4 \\
& - 2\epsilon_r \sigma A_r T_r^4 - k \frac{A_{clegs}}{L_{leg}} (T_r - T_g),
\end{aligned}
\tag{4.42}
$$

where m is the mass of the robot, $c_p^r = 900\frac{J}{kgK}$ is the specific heat of the robot. $\dot{Q}_{gen}''' V_r$ is the internal power generation of the robot estimated at 36W. We assume an emissivity of outer space $\epsilon_{out} = 1$ and a temperature $T_{out} = 2.7K$. Equation (4.42) can be discretized in time using a backward difference to model the evolution of the robot temperature T_r over a sample period Δt, yielding

$$
\begin{aligned}
T_{r,k} = & T_{r,k-1} + \frac{\Delta t}{mc_p^r}(\dot{Q}_{gen,k}''' V_r + G_{n,k} A_r + \epsilon_g \sigma A_r T_{g,k}^4 \\
& + \epsilon_{out} \sigma A_r T_{out,k}^4 - 2\epsilon_r \sigma A_r T_{r,k}^4 \\
& - k\frac{A_{clegs}}{L_{leg}}(T_{r,k} - T_{g,k})).
\end{aligned}
\tag{4.43}
$$

Given a desired nominal operating temperature for the robot $T_{r,nom}$, we define a thermal penalty term, E_{th} as

$$
E_{th} = \beta_1 |T_{r,k} - T_{r,nom}| + \beta_2 (T_g - T_{r,nom})^2,
\tag{4.44}
$$

where $|T_{r,k} - T_{r,nom}|$ penalizes excursion of the robot temperature from the nominal value. This term is particularly relevant over long missions, where the robot temperature has enough time to change. The squared terms added to the cost function penalize in a more direct way paths that require the robot to traverse surfaces with temperatures that differ from the nominal robot temperature. The β_i coefficients are employed to penalize temperature excursions above and below $T_{r,nom}$ in an asymmetrical manner. Here, these parameters are selected as follows:

$$\beta_1 = \begin{cases} 0, & \text{if } (T_{r,k} - T_{r,nom}) > 0 \\ 100, & \text{otherwise;} \end{cases} \tag{4.45}$$

$$\beta_2 = \begin{cases} 0, & \text{if } T_g > T_{r,nom} \\ 0.025, & \text{otherwise.} \end{cases} \tag{4.46}$$

4.4.2 Thermomechanical Cost Function

To develop a physically informed motion plan, a cost function associated with net mechanical and thermal energy consumption is developed. For high-level planning purposes, we treat the ETQuad as a single rigid body moving on the local plane spanned by the unit vectors $\hat{b}_x$ and $\hat{b}_y$. Simplifying assumptions such as minimal side slip and negligible pitch and roll dynamics are considered. This is done to lower computational cost and allow for fast replanning.

The velocity of the center of mass of the robot, B_{cm} is $^N V^{B_{cm}} = v_x \hat{b}_x$, and the robot's angular velocity is given by $^N \omega^B = \omega_z \hat{b}_z$. The leg-terrain interaction is modeled by a longitudinal force $F_x \hat{b}_x$, a lateral force $F_y \hat{b}_y$, which enforces the no side slip velocity and a torque $\tau_z \hat{b}_z$. In addition, dissipative forces and moments include a linear viscous force $F_d = -b v_x \hat{b}_x$ acting at B_{cm} and a linear viscous moment $\tau_d = -b_\tau \omega_z \hat{b}_z$. ETQuad equations of motion in body coordinates are then governed by

$$(F_x - b v_x)\hat{b}_x + F_y \hat{b}_y - mg\hat{n}_z + N\hat{b}_z = m(\dot{v}_x \hat{b}_x + \dot{v}_y \hat{b}_y) \text{ and} \tag{4.47}$$

$$(\tau_z - b_\tau \omega_z)\hat{b}_z = I_{zz} \, ^N \dot{\omega}^B \hat{b}_z, \tag{4.48}$$

where N is the normal force, m is the robot mass, and I_{zz} is the robot's moment of inertia about the $\hat{b}_z$ axis. Therefore, the mechanical power due to the actuators becomes

$$F_x v_x + \tau_z \omega_z = b v_x^2 + m\dot{v}_x v_x + b_\tau \omega_z^2 + I_{zz} \, \dot{\omega}_z \omega_z + mg v_x \hat{n}_z \cdot \hat{b}_x \tag{4.49}$$

We can discretize (4.49) and multiply by the sample period Δt, to obtain the mechanical energetic cost E_{mech} over one sample period, yielding

$$E_{mech} = m|(v_{x,k} - v_{x,k-1})v_{x,k}| + bv_{x,k}^2 \Delta t$$
$$+ I_{zz}|(\omega_{z,k} - \omega_{z,k-1})\omega_{z,k}| + b_r \omega_{z,k}^2 \Delta t + mg|\Delta h|, \quad (4.50)$$

where Δh is the change in terrain elevation, and the absolute values in the expression are used because when the actuators do positive or negative work on the robot, they expend energy.

Equation (4.50) is enhanced with the thermal penalty and cast in the form,

$$E_{cost} = E_{th} + E_{kin} + \beta_3 E_d + \beta_4 E_p, \quad (4.51)$$

where E_{kin} is the cost to accelerate or decelerate the robot, E_d is the cost to overcome frictional forces, E_p is the cost to climb or descend, and E_{th} is a the thermal cost detailed in Section 4.4.1. The mechanical energy costs are

$$E_{kin} = m|(v_{x,k} - v_{x,k-1})v_{x,k}| + I_{zz}|(\omega_{z,k} - \omega_{z,k-1})\omega_{z,k}|, \quad (4.52)$$

$$E_d = bv_{x,k}^2 \Delta t, \quad (4.53)$$

$$E_p = mg|\Delta h|. \quad (4.54)$$

The coefficients β_3 and β_4 are selected as follows:

$$\beta_3 = \begin{cases} 1.0, & \text{if } v_{x,k} > 0 \\ 2.0, & \text{otherwise;} \end{cases} \quad (4.55)$$

$$\beta_4 = \begin{cases} 1.0, & \text{if } \Delta_h > 0 \\ 0.25, & \text{otherwise.} \end{cases} \quad (4.56)$$

For forward propagation, in this application SBMPO employs a single integrator model with control inputs being the robot forward velocity $v_{x,k}$ and angular velocity $\omega_{z,k}$. In particular,

$$
\begin{bmatrix}
X_{G,k+1} \\
Y_{G,k+1} \\
\Psi_{G,k+1}
\end{bmatrix}
=
\begin{bmatrix}
X_{G,k} + v_{x,k}\cos(\Psi_{G,k})T \\
Y_{G,k} + v_{x,k}\sin(\Psi_{G,k})T \\
\Psi_{G,k} + \omega_{z,k}T
\end{bmatrix}
\tag{4.57}
$$

For computational efficiency, in this work the 3D environmental workspace is represented using a 2.5D map generated using the Grid Map library [7]. Thus, motion planning is performed on the plane but the cost model (4.51) queries the map to calculate the cost function, which demands knowledge of the terrain elevation.

Key Steps to Thermally Aware Motion Planning

Referring to Section 2.5, the key steps are as follows.

1. *Sample control space.* In this work, we sample the vehicle forward velocity $v_{x,k}$ and angular velocity $\omega_{z,k}$ using fixed samples.

2. *Generate neighbor nodes.* Use the single integrator model (4.57) with the control samples to determine the neighbors of the current node. Here we employ a sample period, $\Delta t = 7\text{s}$. During this interval, the sampled control inputs are maintained constant and collision checking is performed every 1sec. During this step, the cost to transfer the system from node $n1$ to $n2$ is computed using (4.51).

3. *Compute the heuristic for each node and add the node to the graph.* From the updated robot state, the time to goal $t_{g,k+1}$ is estimated assuming level ground and max forward speed. The heuristic is then calculated using

$$
\begin{aligned}
h(k+1) = {} & m|(v_{x,max} - v_{x,k+1})v_{x,max}| \\
& + b v_{x,max}^2 t_{g,k+1} \\
& + I_{zz}|(\omega_{z,max} - \omega_{z,k+1})\omega_{z,max}|.
\end{aligned}
\tag{4.58}
$$

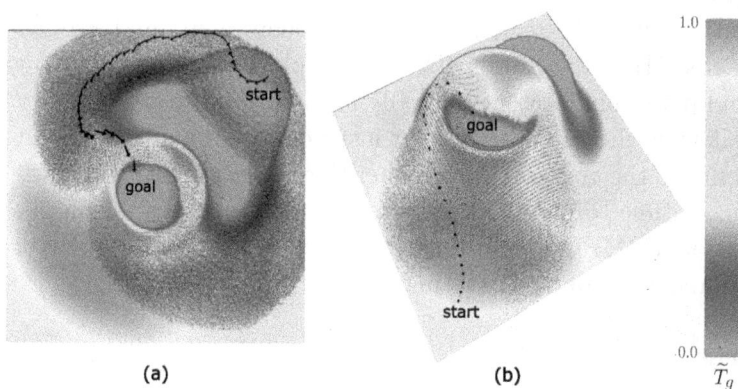

Figure 4.22 (a) Motion planning results with robot starting on a cold region $T_g = 200$K. (b) Motion planning results with robot starting on hot region $T_g = 364.2$K Source: [6].

In the heuristic calculation it is assumed that the ground and robot temperatures are equal to the robot nominal temperature, $T_{r,nom}$.

Then, add the new node to the graph.

Fig. 4.22 illustrates thermally informed motion planning for ETQuad in a crater-like feature exploration mission. In Fig. 4.22(a), the robot starts its mission on a cold region at $(x, y, \psi) = (20\text{m}, 122\text{m}, 5.49\text{rad})$ with a ground temperature $T_g = 200$K. The goal of the robot is located inside the crater within the cold region ($T_g = 200$K) at $(x, y) = (65\text{m}, 65\text{m})$. The motion planner successfully finds a trajectory that takes the robot out of the cold ground early in the mission and proceeds toward the goal in the crater along more moderate temperatures. The plot shows a superposition of the ground temperature and the expansions (black dots) performed by SBMPO while optimizing the vehicle trajectory. The computational time to plan this trajectory was 742ms.

In Fig. 4.22(b), the robot starts its mission at $(x, y, \psi) = (122\text{m}, 20\text{m}, 2.36\text{rad})$ with a ground temperature $T_g = 364.2$K. The goal of the robot is located inside the crater within the cold region at $(x, y) = (65m, 65m)$ and ground temperature $T_g = 200K$. The motion planner guides the robot and determines an energetically efficient route to the goal by following the rim of the crater instead of taking a more direct route to the goal but with higher thermal cost. The computational time to plan this trajectory was 367ms. For this application, SBMPO was integrated with the Grid Map Library of [7] and the Robot Operating System (ROS).

Bibliography

[1] O. Chuy, E. Collins, A. Sharma, and R. Kopinsky, "Using dynamics to consider torque constraints in manipulators planning with heavy loads," *ASME Journal of Dynamic Systems, Measurement, and Control*, 2016.

[2] C. Ordonez, N. Gupta, O. Chuy, and E. Collins, "Momentum based traversal of mobility challenges for autonomous ground vehicles," in *Proceedings of the IEEE Conference on Robotics and Automation*, (Karlsruhe, Germany), May 2013.

[3] N. Gupta, C. Ordonez, and E. G. Collins, "Dynamically feasible, energy efficient motion planning for skid-steered vehicles," *Autonomous Robots*, vol. 41, pp. 453–471, Feb 2017.

[4] G. Francis, E. Collins, O. Chuy, and A. Sharma, "Sampling-based trajectory generation for autonomous spacecraft rendezvous and docking," in *AIAA Guidance, Navigation, and Control Conference*, (Boston, MA), 8 2013.

[5] G. D. Francis, *Novel Applications of Sampling-Based Model Predictive Optimization*. PhD thesis, Florida State University, Tallahassee, FL, October 2019.

[6] C. Ordonez, J. Ordonez, J. Boylan, D. Vazquez, and J. Clark, "Thermally informed motion planning to

enhance mission endurance of mobile robots," in *Proceedings of the 8th Thermal and Fluids Engineering Conference*, (College Park, MD), March 2023.

[7] P. Fankhauser and M. Hutter, "A Universal Grid Map Library: Implementation and Use Case for Rough Terrain Navigation," in *Robot Operating System (ROS) – The Complete Reference (Volume 1)* (A. Koubaa, ed.), ch. 5, Springer, 2016.

[8] B. Siciliano, B. Sciavicco, L. Villani, and G. Oriolo, *Robotics Modelling, Planning, and Control.* Hemisphere Publishing Corporation, 2009.

[9] W. Yu, O. Chuy, E. G. Collins, and P. Hollis, "Analysis and experimental verification for dynamic modeling of a skid-steered wheeled vehicle," *IEEE Transactions on Robotics*, vol. 26, pp. 340–353, 2010.

[10] C. Ordonez, N. Gupta, B. Reese, N. Seegmiller, A. Kelly, and E. Collins, "Learning of skid-steered kinematic and dynamic models for motion planning," *Robotics and Autonomous Systems*, vol. 95, pp. 207–221, 2017.

CHAPTER 5

Learning Models and Heuristics: Current and Future Work

SBMPO is dependent upon the existence of models, which may be quite complex and may also change over time. In many cases, these models are difficult or impractical to develop analytically. Hence, for non-fragile applications of SBMPO, planning needs to be accompanied by algorithms that enable model learning. Likewise, the computational performance of SBMPO is dependent upon the non-conservativeness of the employed heuristics. It is not only desirable that heuristics be optimistic, but they also need to be non-conservative to ensure fast planning.

Section 5.1 presents results that illustrate the learning of dynamic and power models, and provides directions for future research. Section 5.2 presents a methodology for learning heuristics and provides an example that motivates the use of multiple heuristics. It also presents directions for future research.

The results in this section are preliminary results. Some of the results and discussion is new and some has appeared in conference papers. The results related to a climbing robot

DOI: 10.1201/9781003623830-5

have appeared in a short journal paper. A primary reason for this presentation is to motivate future work as the authors are convinced that there is much work to be done in learning both models and heuristics.

5.1 LEARNING MODELS

The first two results below consider learning specialized dynamics of a more complex system. In particular, Section 5.1.1 starts by considering learning of a slip model that can be used to enhance the kinematic model for a skid-steered vehicle. Then, it considers learning the frictional term in the dynamic model of a skid-steered vehicle.

The next two results consider learning the complete dynamics for complex robot systems. Section 5.1.2 considers learning dynamic and power models for the LLAMA walking robot whereas Section 5.1.3 considers learning the dynamic model of a climbing robot.

5.1.1 Learning Skid-Steered Dynamic Models for Energy-Optimal Motion Planning

Planning with skid-steered vehicles is challenging because of the complex interaction between the wheels/tracks and the terrain [1]. Due to their geometry and steering approach, these vehicles experience large amounts of forward, lateral, and angular slip [2]. As illustrated in Fig. 5.1, actual and commanded vehicle trajectories can differ significantly because the vehicle motion is dependent on the commanded velocities, the robot payload, and the surface properties. The difference between actual and planned trajectories has major negative implications. First, the replanning rates will need to be increased to add some form of robustness. Second, the robot's actual trajectory can result in immobilization because it results in a collision or an attempt to traverse a patch of impassable terrain. Third, the predicted cost (e.g., energy) of vehicle motion can be very different from the actual cost. To remedy this,

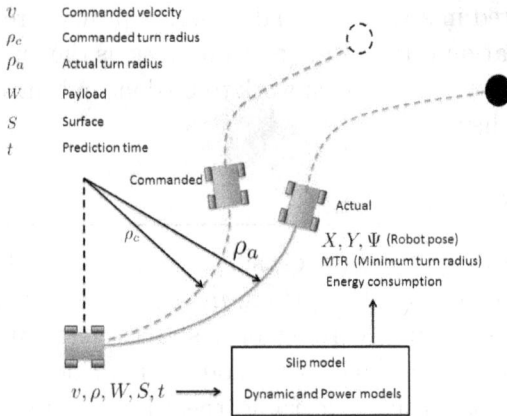

Figure 5.1   Difference between actual and commanded trajectories for skid-steered vehicles. This difference varies as a function of vehicle commands, surface properties, and vehicle payload. Learning of the slip-enhanced kinematics and the vehicle dynamics can compensate for this difference and thus help reduce replanning rates.

both the kinematic and dynamic vehicle models need to be adapted to the surface being traversed.

Slip Enhanced Kinematic Model

To minimize pose errors, a slip-enhanced kinematic model can be calibrated using the Integrated Predictive Error Minimization (IPEM) algorithm [1, 3]. The approach models the slip-enhanced vehicle kinematics as

$$v_{lon} = v_c + \delta v_{lon}, \tag{5.1}$$

$$v_{lat} = \delta v_{lat}, \tag{5.2}$$

$$\Omega = \Omega_c + \delta\Omega, \tag{5.3}$$

where v_{lon}, v_{lat}, and Ω represent the longitudinal, lateral, and angular velocities of the vehicle, respectively v_c and Ω_c are the commanded velocities, and δv_{lon}, δv_{lat} and $\delta\Omega$ correspond to

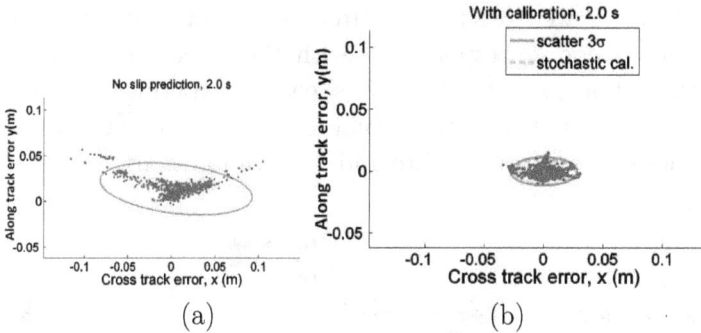

Figure 5.2 Along and cross-track errors before (a) and after (b) calibration during a diagnostic trajectory on an asphalt surface. Notice that the IPEM method calibrates deterministic and stochastic errors.

the longitudinal, lateral, and angular slip, respectively which are modeled as the following polynomial functions of the velocity commands:

$$\delta v_{lon} = a_{11}v_c + a_{12}\Omega_c + a_{13}v_c\Omega_c, \tag{5.4}$$

$$\delta v_{lat} = a_{21}v_c + a_{22}\Omega_c + a_{23}v_c\Omega_c, \tag{5.5}$$

$$\delta\Omega = a_{31}v_c + a_{32}\Omega_c + a_{33}v_c\Omega_c. \tag{5.6}$$

IPEM calibrates the coefficients a_{ij} as the robot moves in the terrain of interest, and it does so in an online and real-time fashion, using the first integral of the differential equation being calibrated [3]. When calibrating the slip-enhanced kinematic model, it is beneficial, although not required, to start the calibration with a nominal model (e.g., the result of a previous calibration).

A typical, slip model calibration result for the robot of Fig. 4.15 moving on an asphalt surface is shown in Fig. 5.2, which compares along and cross-track errors for the uncalibrated and calibrated trajectories. In this figure, the errors are computed as the difference between the predicted robot pose by the slip-enhanced kinematic model and the actual pose estimated using visual odometry. The errors are estimated using a prediction horizon of 2s. Notice that the cross and along

track errors are close to zero after calibration and that there
is also a good correspondence with the three sigma ellipses
in the calibrated case. Fig. 5.3, shows an experimental result
on asphalt, comparing the actual robot trajectory against the
predicted trajectories before and after calibration.

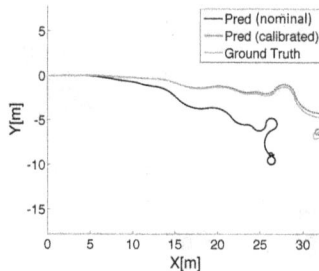

Figure 5.3 Comparison, on an asphalt surface, of the actual
robot trajectory against the predicted trajectories before and
after slip calibration using IPEM.

Friction Model

As discussed in Section 4.3, the energetic cost incurred by a
skid-steered vehicle when moving from one node to the next
is given by $E_k = P_k T$, where T is the sample period and
P_k is the required power, which is a function of the inner
and outer wheel torques and wheel velocities (see Fig. 4.18).
As the surface changes, the frictional term $C(q, \dot{q})$ in (4.1) or
equivalently the torque vs. turn radius curves (as in Fig. 4.16)
can change significantly. One solution to this problem consists
in learning approximations to these curves. The methodology
developed in [1] employs a Minimum Resource Allocation Network (MRAN) [4], which learns representations of the wheel
torques for each side of the vehicle using two different (one per
robot side) Radial Basis Function (RBF) neural networks. The
MRAN algorithm allows the network to automatically change
its size to adjust to changes in the system being learned. In
addition, thanks to the utilization of an Extended Kalman

Filter (EKF) formulation, the network adaptation is performed online and in real time.

Fig. 5.4, shows learned curves for the skid-steered vehicle of Fig 4.15 when it traverses an unknown asphalt surface.
The networks were initialized using the torque vs. turn radius
data obtained from the terramechanic model of the vehicle
on a vinyl surface [5]. Besides aiding in the prediction of energetic cost, these torque functions generate estimates of the
minimum turn radius (MTR), which is the sharpest turn that
the vehicle can be commanded without incurring torque saturation and possibly uncontrollable motion [6].

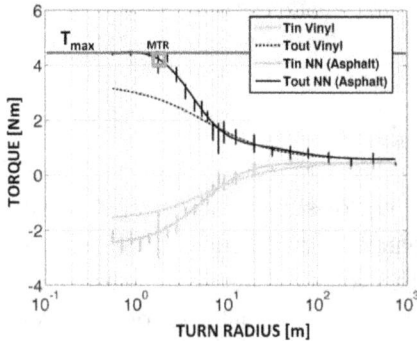

Figure 5.4 Learned torque vs. turn radius representations for
an asphalt surface using the MRAN algorithm. These curves
also generate predictions of the minimum turn radius (MTR),
which is the sharpest turn that the vehicle can be commanded
without incurring in torque saturation. [1] C. Ordonez, N.
Gupta, B. Reese, N. Seegmiller, A. Kelly, and E. Collins,
"Learning of skid-steered kinematic and dynamic models for
motion planning,," Robotics and Autonomous Systems, vol.
95, pp. 207–221, 2017.

Overall Modeling Approach

The overall methodology for learning skid-steered slip and dynamic models for motion planning is summarized in the block
diagram of Fig. 5.5. A key feature of the methodology is that

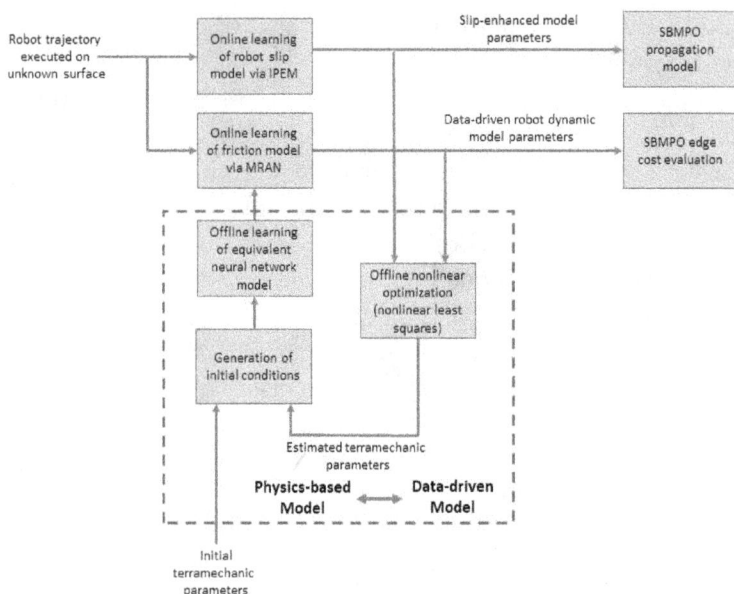

Figure 5.5 Learning of slip-enhanced and dynamic models for skid-steered vehicles. The learned models are used by SBMPO to propagate the vehicle state and to estimate the energetic cost between nodes. The methodology allows for switching between physics-based models and data-driven models, which is important when there are payload changes [1].

it allows for switching between physics-based and data-driven models. This is not a requirement but it is advisable when there are changes in the vehicle payload [1].

Fig. 5.6 illustrates the benefits of energy efficient planning when employing the learned slip and dynamic models of the skid-steered vehicle of Fig. 4.19. By learning these models, the vehicle was able to plan and execute (without replanning) an energetically efficient trajectory.

5.1.2 Learning Dynamic and Power Models of the LLAMA Legged Robot for Energy Optimal Planning

The LLAMA is a 45kg, 60cm tall quadrupedal robot designed to move at human speeds and carry a payload of sensors and

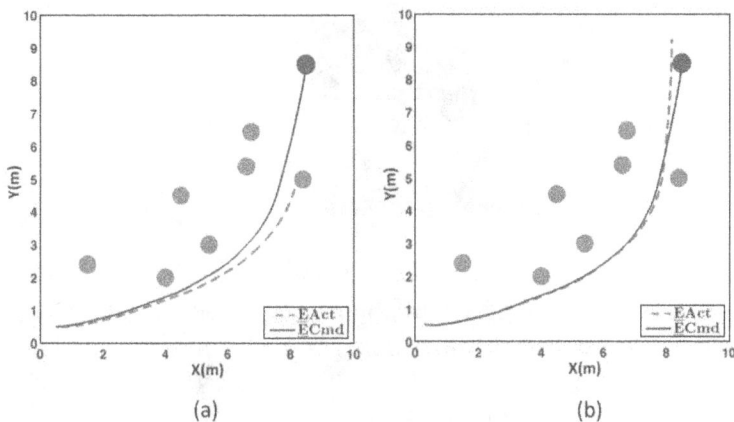

Figure 5.6 Learning of slip and dynamic models for the skid-steered vehicle of Fig. 4.19. The learned models are used for energy efficient motion planning on a concrete surface. (a) SBMPO using the vehicle models for asphalt. Notice the collision with the obstacle. (b) SBMPO using the adapted slip models for the concrete surface. The predicted mechanical energy was 622J and the actual energy was 521J.

equipment suitable for use in human-robot teaming [7]. As this robot was being constructed and undergoing component tests, a simulation environment (Fig. 5.7) was created using detailed electrical and CAD schematics to understand the motion dynamics and power consumption of this platform. The simulation environment was essential to use as a data collection tool as simplified analytical models were unable to replicate the fidelity of kinematic and dynamic legged motion.

The complexity in predicting whole-body motion of legged robots encourages use of tools such as motion primitives [8] or a hierarchical control structure [9]. We chose to utilize learned models derived from data collected in motion capture simulations. After testing several neural network algorithms, it was found that a simple feed-forward neural network had sufficient capacity to act as a motion model. The simple nature of the network was also easy to evaluate and cost very little

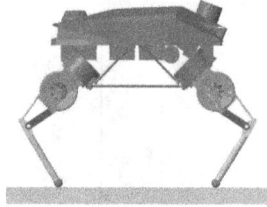

Figure 5.7 The LLAMA platform simulated in the MATLAB®
Simscape Multibody software. Special care was taken to reflect
LLAMA's asymmetric 5-bar linkages, motor models, and low-
level controllers.

computing time to evaluate. This learned motion model serves
as the vehicles' kinematic and power models, converting control inputs to velocity and power consumption estimates.

Another important consideration when planning for the
LLAMA platform was the employed gait, a cyclic pattern of
leg motions used to maneuver and maintain stable locomotion. This type of locomotion requires a fundamental shift in
planning methodology from representing trajectories in terms
of time to representing them in terms of discrete strides (the
completion of one set of cyclical leg motions). During the
execution of a single stride, internal body motions (such as
roll, pitch, yaw) adds significant volatility to robot pose. By
limiting the data to stride, it has been found [10] that data
is naturally filtered while maintaining relevant information to
determine an accurate motion model.

Gait-Based Adaptation of SBMPO

The dependence on stride changes the index k in Fig. 2.4
from a general time index into a stride index. This change
also modifies the cost function and optimistic heuristic function. Distance optimal heuristics were adapted to be a Euclidean distance measured in attainable stride lengths between
a considered node and the goal. The distance-based cost was

calculated by multiplying the number of strides taken by the distance traveled by each stride.

In the case of energy optimal planning, let D denote the Euclidean distance between the current node and the goal and let t denote the time it takes for the robot to travel D at a fixed stride frequency resulting in a forward velocity $v_{x,k}$, such that $t = \frac{D}{v_{x,k}}$. The optimistic heuristic is then given by the energy expended, i.e., $E = tP_\infty$, where P_∞ denotes the power consumed in straight line motion.

Model outputs of center of mass (COM) linear velocities $v_{x,k}$ and $v_{y,k}$ and body angular velocity ω are propagated into pose estimates X, Y, and θ by stride through the single integrator model,

$$X_{k+1} = X_k + (v_{x,k}cos(\theta_k) - v_{y,k}sin(\theta_k))T, \quad (5.7)$$
$$Y_{k+1} = Y_k + (v_{x,k}sin(\theta_k) + v_{y,k}cos(\theta_k))T, \quad (5.8)$$
$$\theta_{k+1} = \theta_k + \omega_k T. \quad (5.9)$$

Data-Driven Approach

The neural networks used in the final models were feedforward networks that have two hidden layers. The network was structured to have 40 hidden neurons in the first layer and 3 in the second layer with the output layer also having 3 neurons. Inputs to the network are the control parameters f_k (stride frequency) and DY_k (turn gain, a lengthened leg sweep induced asymmetrically to produce a turning behavior). The outputs of the model were forward, $v_{x,k}$, and lateral, $v_{y,k}$, linear velocities of the COM, angular velocity of the robot body ω_k, and power consumption P_k.

To generate data for the neural network, a series of simulations were executed [11]. These simulations spanned the stable range of control inputs, providing the necessary breadth of power and motion data for the neural network. In this manner, the neural network was constructed entirely within the stability bounds of the vehicle.

Data on control inputs was collected in simulation per stride, which was used to train the neural networks of

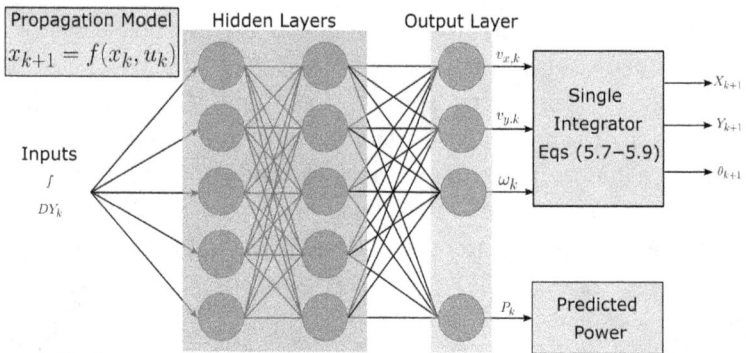

Figure 5.8 The propagation model converts control inputs into outputs and is used within SBMPO to generate the predicted next state. The central component of this is a neural network which learns the kinematics, this is coupled with a single integrator model (5.7)–(5.9). To generate energy efficient trajectories, a second network with identical structure is trained on power consumption data. The key outputs of the two-network propagation model is robot pose (position, orientation) and power. Note that for simplicity, in this figure only one network is shown.

Fig. 5.8 to generate velocity and power consumption predictions. Data collection, testing, and validation of the motion planning were carried out in the MATLAB Simscape Multibody software [11]. A scenario map of obstacles was loaded into SBMPO which returns a series of stride-by-stride control instructions f and DY. Execution of these instructions was completed on the simulated robot, which was constrained to only accept new commanded controls once the stride cycle had completed.

Several scenarios in vegetated outdoor environments were tested in simulation and in each case the energy efficient mode of optimization resulted in reduced energy consumption as fewer high-energy maneuvers were used at the cost of increased traversal distance. Fig. 5.9 illustrates one of the scenarios

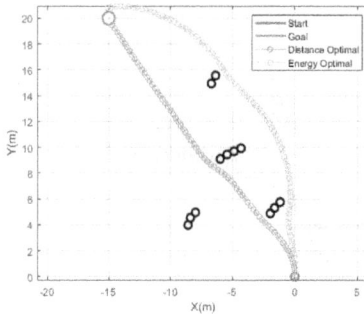

Figure 5.9 Distance-based (black) and Energy-based (gray) trajectories. Each trajectory moves from the start point (0,0) to the goal (-15m,20m).

considered and shows the two SBMPO trajectories computed using the distance and energy cost functions.

The corresponding metrics are given in Table 5.1 for simulations performed in an AMD TR 1950X processor (3.9GHz) with 128GB RAM. For another scenario, Table 5.2 shows the model, predicted distance and energy costs as compared to the results of executing the computed trajectory in a physics-based simulation. While actual energy and distance traversals were higher in simulation, this increase stemmed from the robot experiencing momentum effects during rapid changes in maneuvers that were not modeled in data collection.

Table 5.1 Metrics Associated with the SBMPO Trajectories of Fig. 5.9

Cost Function	Energy	Distance
Computation Time (sec)	0.26	0.021
Strides	84	71
Distance (m)	28.9	24.7
Energy (kJ)	35.7	43.1

Table 5.2 Metrics Associated with SBMPO Trajectories on LLAMA

Cost Function	Energy		Distance	
Model	Neural Net	Simscape	Neural Net	Simscape
Distance (m)	23.7	24.0	22.9	23.0
Energy (kJ)	27.0	30.5	41.5	44.8

5.1.3 Learning a Dynamic Model of the TAILS Climbing Robot for Distance-Optimal Planning

The dynamic locomotion characteristics developed for the LLAMA were transitioned for use on the TAILS platform [12]. TAILS is one of the fastest climbing robots in the world [13] capable of climbing over 1.5 body-lengths per second. This speed is made possible by animal inspired Full-Goldman climbing dynamics. This robot exhibits similar motion constraints to LLAMA where it has unusual (both holonomic and non-holonomic) motion, and stride-based (discrete) state estimation and control authority.

While these challenges are inherent to motion planning on most legged platforms, dynamic vertical climbing adds an additional series of very unusual foot contact limitations, alterations in body geometry throughout a stride heading angle constraint, and limited lateral motion (detailed in Section: 5.1.3). Without these constraints, the robot is unable to maintain a stable foothold and disengages from the climbing surface (resulting in catastrophic failure). Navigation within vertical climbing has been limited to quasi-static foot-step planning [14, 15, 16], primarily because of the limited maneuverability of dynamic climbers.

Defining TAILS Motion

Much of the difficulty in the motion modeling of TAILS stems from heading angle constraints; the robot must always be

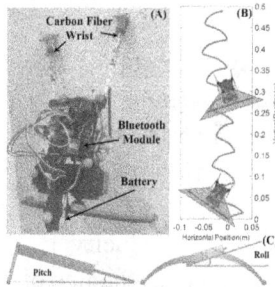

Figure 5.10 (a) TAILS, one of the fastest climbing robots in the world. (b) A typical COM trajectory during upward climbing. The Full-Goldman climbing dynamic features a pendular swing with each toe-hold. Bounding boxes defining the robot geometry are comprised of two parts with a static box defining the rigid body and a variable-sized trapezoid defining the tail region. (c) Two actuated tails at the bottom of the robot control body pitch and roll. Source: [13]. M. P. Austin, M. Y. Harper, J. M. Brown, E. G. Collins, and J. E. Clark, "Navigation for legged mobility: Dynamic climbing," IEEE Transactions on Robotics, 2019.

facing upward in order to maintain attachment. Thus constrained, maneuverability can only be achieved via fast upward climbing, slow downward climbing, and diagonal strafing. Upward climbing can be accomplished within a range of velocities that preserve natural pendular dynamics. Downward climbing is more difficult as the robot descends deliberately to mitigate loss of stability and attachment; this results in much reduced speed. Finally, as orientation and surface attachment constraints bar direct lateral motion, a slight diagonal motion (strafing) can be induced at reduced speed from purely vertical climbing.

The body geometry of the robot also varies with the climbing speed, induced behavior, and prescribed posture. In previous results it has been found that the body swing magnitude is much higher at low retraction frequencies and is attenuated at high frequencies, which causes the effective lateral size of the body to increase or decrease [12].

Due to the difficulty in mapping the control parameters to anticipated motion, a learning approach similar to that for LLAMA was employed. A series of motion-data collection experiments were conducted to determine the prediction models relating the control parameters (f, Offset, and Pitch) to upward velocity V_Z and lateral velocity V_X [13]. Similar to the LLAMA, f denotes the stride frequency of climbing. The other two control terms Offset and Pitch are related to body roll induced by the tail offset and the angle maintained respective to the climbing surface (Pitch).

Three separate feed-forward neural networks were employed as the motion models with each model trained on a specific body pitch, reducing the complexity and data requirements. This modeling paradigm was selected for ease of use in online adaptation and its capacity to account for much of the motion non-linearity.

This data was processed using a stride filter [10] and models were trained on extracted data. Due to limitations of the motion capture space, most experiments were comprised of only one type of motion (i.e., strafing right, straight down, etc.) and failed to capture transition effects between two different motions. This resulted in some unmodeled behavior during transitions between high-speed vertical to low-speed downward maneuvers [13]. However, these motion models were sufficiently accurate for successful trajectory planning.

Directional Heuristic, Cost, and Obstacle Definitions

A new cost function and optimistic heuristic function were required because of the unique motion characteristics of TAILS. Different maneuvers incur in higher distance costs than others (Fig. 5.11), thus, a cost and heuristic function were created to account for the relative ease of vertical mobility, slower downward motion, and the increased costs of horizontal movement.

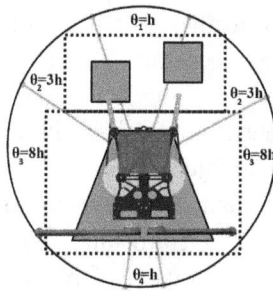

Figure 5.11 TAILS mapping of directional heuristic scaling. The region defined by θ_1 is within the limits of robot strafing and can be reasonably approximated by Euclidean distance h. Attaining a point outside that region cannot be approximated by simple Euclidean distance due to the difficulty of navigation; thus, an inflation term is added to the heuristic as the robot requires more distance and time to traverse horizontally due to the fixed heading angle constraint. Source: [13]

A distance-based heuristic was defined using Euclidean distance between the COM and the goal. This heuristic was then inflated using a scalar based on direction of the goal region relative to the robot (shown in Fig. 5.11). This directional inflation factor is key in rapid computation of trajectories for the TAILS planner. An example trajectory was computed for a goal coordinate defined at 2m to the right of the robot. The computation time using a naive heuristic of Euclidean distance took 187.3s and explored 192,711 nodes in the planning space. The adjusted heuristic function was able to reduce the planning time to 0.36s with exploration of only 3,197 nodes.

The cost function was similarly adjusted to reflect the difficulty in certain types of motion. Distances traveled differed greatly in downward and lateral maneuvers. These costs were determined through experimental data collection.

A stability penalty was also introduced. This was a penalty added to explorations that sampled maneuvers with a higher chance of attachment failure, which can cause the robot to fall off the wall. This focused the search on the more stable motions but allowed the algorithm to explore less stable regimes if it was necessary (due to body geometry constraints).

A final change in the algorithm allowed the robot to be treated as two separate objects (shown in dotted rectangles in Fig. 5.11), each with their own constraints and interactions. The feet are treated in a traditional manner with obstacle detection preventing exploration of regions overlapping with the defined foot-area. The body is defined with a 3D constraint and is allowed to interpret when an obstacle does not constrain motion. For example, a hole in the wall is not a constraint for the body but will stop the feet from attaching.

Motion Planning Results on the TAILS Physical Robot

Physical validation was conducted using a VICON system to track the position of the robot and see how well the planner and model could perform. The planned trajectories for four cases (solid gray lines in Fig. 5.12) were tested in a purely feed-forward manner on TAILS; no tracking controller was employed to ensure that the planned trajectories were followed. The experimental path is shown as the dotted lines using the stride-based COM path filtering described in Section 5.1.3. Final robot positions from other trials are shown as black circles.

With the improved heuristic and cost function, unique coupled bounding box obstacle treatment, and learned motion models, SBMPO was able to generate achievable trajectories for each of the test cases. While high-volatility transients impacted performance in lateral motions, cases without significant lateral motions (Fig. 5.12 (a) and (c)) saw little deviation. In all cases, the planner was successfully able to compute a viable trajectory.

Continued work will be done to model and better understand transient behaviors as the robot transitions from one maneuver to another. The high volatility seen in the lateral experiments could likely be mitigated by a combination of

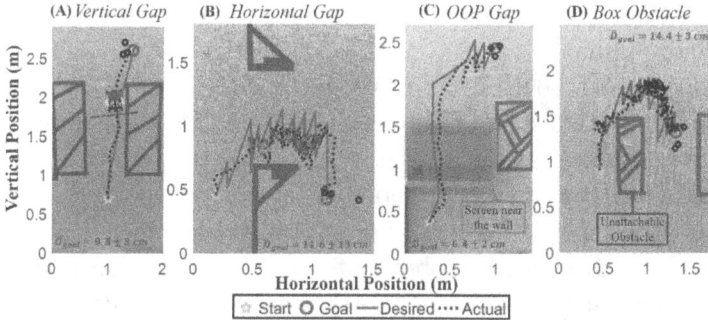

Figure 5.12 Resulting motions from executing SBMPO generated trajectories optimizing distance-based navigation. The SBPMO desired path (gray) connected the start node (star) to the goal node (dark black circle). The robot's swing-filtered body path is shown in black with final positions from three trials in black circles. The average and standard deviation distance from the goal is presented in dark black text. The three versions of the narrow gap problem are shown in (a) through (c), and (d) shows the discrete obstacle. Rapid changes between strafing and downward climbing show the planner trying to generate lateral movement while maintaining the nonholonomic heading constraint. Source: [13].

re-planning and higher fidelity models. Large-scale planning will also be performed to enable trajectory planning for climbing large buildings.

5.1.4 Future Research Directions in Learning Models

While neural networks are powerful tools in processing of data, other algorithms exist which are being incorporated into the SBMPO model-building lexicon. A general data science toolkit integration is planned to allow the best performing models (XGBoost [17], Light GBM [18], and Long Short Term Memory [19]) to operate within the SBMPO framework. Expansion to all SkLearn-[20]-based libraries [20] and GPU support will allow these models to be trained and adapted online in a more efficient fashion.

Learned models of more complex robots need to account for information as a time series and LSTMs have proven to be capable in simulation tests at predicting transient effects which are difficult to model using current tools. Pairing time-series specific models with a parallel implementation of SBMPO [21] gives the robot sufficient time to replan trajectories or track back to the original path after deviation. Additional work is being undertaken to streamline the parallel algorithm to share the graph between processes to lower the memory footprint of the algorithm.

Further, as these algorithms are implemented for wide-area autonomy tasks in unstructured environments, there is a need to account for uncertainty and cases where a single heuristic may not be sufficient. To account for these complexities, an extension of SBMPO will incorporate A-MHA* [22] and probability funnels [23]. These tools will allow combinations of diverse heuristic functions and guarantee safe operation in the presence of uncertainty.

5.2 LEARNING HEURISTICS

Section 5.2.1 concerns learning heuristics for a model predictive control problem. Though not yet accomplished, the general methodology can be extended to cost functions, for example distance, time, and energy. Section 5.2.2 does not consider explicit learning of heuristics, but implicit learning when a suite of heuristics is being considered. The purpose of that section is to motivate the need for the development of SBMPO/MHA*.

5.2.1 Learning Heuristics for Model Predictive Control

A common cost function that is employed in Model Predictive Control (MPC) [24] is given by

$$\sum_{i=0}^{N-1} ||r_{k+i+1} - y_{k+i+1}||^2, \tag{5.10}$$

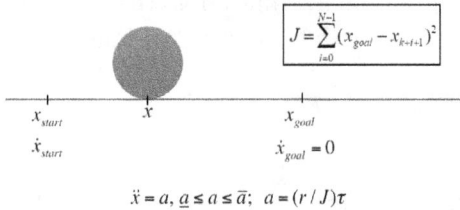

$$J = \sum_{i=0}^{N-1} (x_{goal} - x_{k+i+1})^2$$

x_{start} x x_{goal}

$\dot{x}_{start}$ $\dot{x}_{goal} = 0$

$$\ddot{x} = a, \ \underline{a} \le a \le \bar{a}; \quad a = (r/J)\tau$$

Figure 5.13   Simple double integrator system used to plan a trajectory from start to goal using the traditional MPC minimum deviation cost function.

where r_j is the reference input at time j and y_j is the output at time j. In this cost function the edge cost from j to $j+1$ is $||r_{j+1} - y_{j+1}||^2$. However, determining heuristics associated with this cost function is very difficult, even for simple dynamic systems, a problem exacerbated for nonlinear and/or complex dynamic systems.

Fig. 5.13 presents a planning scenario for the double integrator system discussed in Chapter 3. In this simulation, the start was $(x_{start} = 0, \dot{x}_{start} = 0)$, the goal was $(x_{goal} = 10m, \dot{x}_{goal} = 0)$, and the acceleration constraints were $-1\frac{m}{s^2} \le \ddot{x} \le 1\frac{m}{s^2}$. The branchout factor was 7, grid resolution $0.01m$ in x and $0.001\frac{m}{s}$ in $\dot{x}$, and the sample period was $T = 0.1s$. Using the processor described in Section 3.1, Table 5.3 compares computation times as a function of the different cost functions and heuristics.

While SBMPO was able to solve the planning problem for the double integrator using the MPC cost function, the computational time increased significantly. This is due to the usage of the highly optimistic and trivial heuristic $H = 0$. To remedy this, it is possible to learn heuristics through extensive offline simulations as a way to provide online computational speed that can be used with the MPC function or with other cost functions for which the cost-to-goal is difficult to determine or cannot be derived analytically [25].

Table 5.3 Computational Times for the Double Integrator Problem of Fig. 5.13 with Different Cost Functions and Heuristics

Cost Function	Heuristic	Computation Time [sec]
MPC	0	3.71
Distance	Velocity aware (Sec. 3.3)	0.074
Time	Minimum time (Sec. 3.1)	0.062

Fig. 5.14, depicts the general methodology process, which consists in learning a function to map a vector comprised of input and goal states to the minimum cumulative cost to move optimally from start to goal. This function is approximated using a neural network trained via extensive offline simulations using a batch of randomly selected input and output states. To generate the batch of optimized trajectories, SBMPO is run offline with the trivial heuristic $H = 0$. As an illustrative example, consider the double pendulum of Fig. 5.15. In this

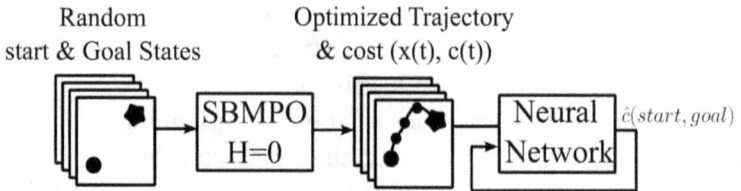

Figure 5.14 Methodology to learn heuristics. Extensive optimal trajectories are simulated offline using randomized start/goal states and the trivial heuristic $H = 0$. A neural network is trained to predict the cost-to-goal c.

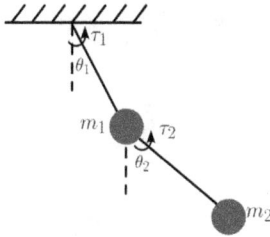

Figure 5.15 Double pendulum system used to illustrate the learning of a heuristic suitable for motion planning with the MPC cost function (5.11).

case, a heuristic H_1, not necessarily optimistic, was learned to speed up SBMPO planning for the MPC cost function

$$\sum_{i=1}^{N}(x_2(i) - x_{2,goal})^2 + (y_2(i) - y_{2,goal})^2, \qquad (5.11)$$

where the prediction horizon N was set to 18 and (x_2, y_2) represents the position of the mass m_2.

The selected neural network architecture for this cost-to-goal learning problem consists of a single hidden layer with a sigmoidal activation function. Using the Levenberg-Marquardt algorithm [26], the neural network was retrained for hidden layer sizes between 3 and 30.

Fig. 5.16(a) compares the cost of various optimal trajectories against the SBMPO trajectory cost predicted with the learned heuristic H_1, which is not necessarily optimistic. Notice the good correspondence between the two, indicating that the learned H_1 is effective in achieving optimal or near-optimal trajectories. In addition, Fig. 5.16(b) compares the computational time required to obtain SBMPO trajectories with and without the learned heuristic. Without the cost-to-goal estimate, the longest computation out of all the trajectories takes 356s. When using H_1, the longest computation requires only 2.0s. The cost-to-goal estimate also improves the median

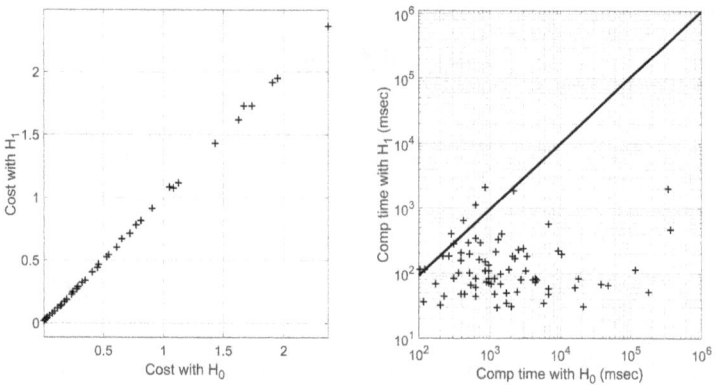

Figure 5.16 (a) Comparison of actual vs. learned cost-to-goal for diverse trajectories of the double pendulum system of Fig. 5.15. Data points in the region $(y < x)$ represent an improvement in optimality when employing the learned cost-to-goal estimate while points in the region where $(y > x)$ represent a decrease in optimality. (b) Comparison of computational times for trajectory planning using the zero heuristic H_0 vs. the learned heuristic H_1. Data point is the regions where $(y < x)$ represent an improvement in computational time when employing the learned cost-to-goal estimate while points in the region where $(y > x)$ represent a slower computational time. Notice the decrease in computational time when using H_1. Without the cost-to-goal estimate, the longest computation took 356s. When using H_1, the longest computation was reduced to 2.0s. The cost-to-goal estimate also improved the median computational time from 1095ms to 84ms.

computational time from 1095ms to 84ms. For further details, refer to [25].

5.2.2 Need for a Multiple Heuristic SBMPO Framework

In realistic long-range outdoor applications, mobile robots are expected to encounter diverse terrain surfaces such as asphalt, grass, dirt, snow, sand, and water, to cite a few. Furthermore, within a single mission, a robot may be required to traverse a large area comprising a combination of such terrains. Due to the long traveling distances and the varied energetic locomotion costs required to traverse these surfaces, it can become computationally expensive to find optimal trajectories. To illustrate this point, here we focus on energy efficient motion planning in multi-terrain environments. However, the issues and concepts being discussed are relevant to other motion planning applications that have multiple heuristics. One similar example is energy efficient planning in environments that have various slopes. Multiple heuristics also occur when a robot has various dynamic models. This situation can occur in wheeled or tracked vehicles that have different gear settings or legged robots that have various gaits.

Building upon the results of Section 4.3, let us consider the skid-steered vehicle shown in Fig. 5.17. This robot is trying to reach the goal in an energy efficient manner while having to traverse three surfaces, each with different torque and power requirements. In this scenario, an optimistic heuristic can be chosen to be the one corresponding to the least resistant terrain. That is, the terrain that demands the least amount of torque τ_∞ and power P_∞ from the robot motors under the assumption of rectilinear motion at constant speed.

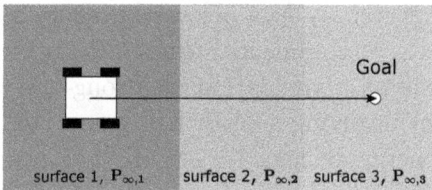

surface 1, $P_{\infty,1}$ surface 2, $P_{\infty,2}$ surface 3, $P_{\infty,3}$

Figure 5.17 Skid-steered vehicle planning an energy efficient trajectory in a multi-terrain environment. Note that this simple planning problem involves multiple heuristics. The heuristic actually used can determine the computational efficiency of SBMPO.

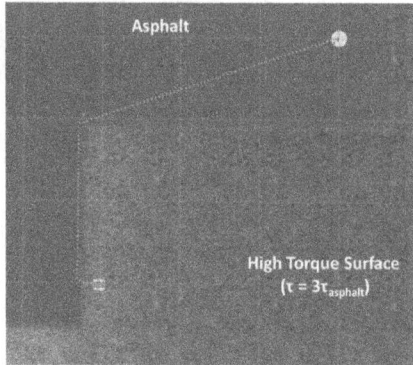

Figure 5.18 Comparison of distance and energy efficient trajectories on a multi-terrain environment composed of asphalt and grass surfaces. Notice that the energy efficient trajectory prefers a longer path in order to avoid as much as possible the energetically demanding grass surface.

The optimistic heuristic H can be computed as

$$H = P_\infty \Delta t, \tag{5.12}$$

where

$$P_\infty = \min\{P_{\infty,1}, P_{\infty,2}, P_{\infty,3}\}, \tag{5.13}$$

and Δt is the required time to travel from the current state to the goal. However, as seen below, this choice of a heuristic may lead to long computational times.

Fig. 5.18 shows a simulation of a long-range mission in an environment involving asphalt and grass. For comparison, distance and energy efficient trajectories are included. Notice that the energy efficient trajectory tries to avoid the energetically costly grass surface even though this results in a longer path. In this simulation, the required torques for curvilinear motion on grass were assumed to be three times as large as those for asphalt. The caveat is the increased computational time. Table 5.4 provides quantitative results for this scenario using a 1.8GHz dual core Intel Core i5 GPU.

Table 5.4 Quantitative Motion Planning Results on the Multi-terrain Environment of Fig. 5.18.

Optimization Problem	Distance Traveled, m	Energy Consumed, kJ	Computation Time, s
Minimum Distance	44.5	15.6	21.7
Minimum Energy	55.5	9.5	384.3

A second scenario is presented in Fig. 5.19, where a large obstacle is placed within the vehicle start location and the desired goal. The obstacle is surrounded by sand, and the goal is located in the asphalt region. The figure compares energy efficient planning results using optimistic (asphalt heuristic) and non-optimistic (sand heuristic) energy heuristics. There is a significant difference in computation time but only a relatively small improvement in energy optimality. Fig. 5.20 explains the difference in computation times. SBMPO had to expand a large number of nodes when using the optimistic heuristic.

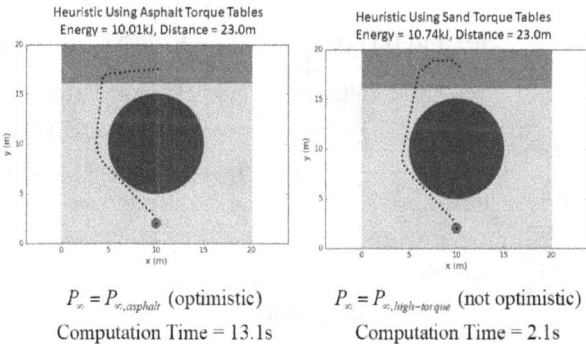

Heuristic Using Asphalt Torque Tables
Energy = 10.01kJ, Distance = 23.0m

Heuristic Using Sand Torque Tables
Energy = 10.74kJ, Distance = 23.0m

$P_\infty = P_{\infty, asphalt}$ (optimistic)
Computation Time = 13.1s

$P_\infty = P_{\infty, high-torque}$ (not optimistic)
Computation Time = 2.1s

Figure 5.19 Energy efficient trajectories in a multi-terrain environment comprised of sand and asphalt.

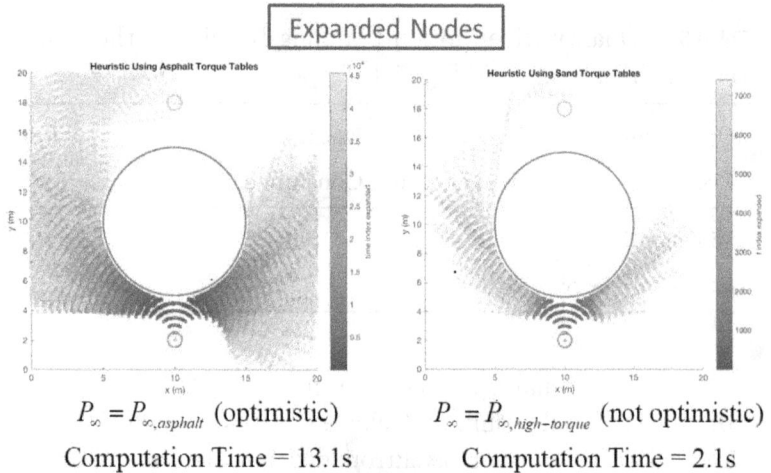

$$P_\infty = P_{\infty,asphalt} \text{ (optimistic)}$$
Computation Time = 13.1s

$$P_\infty = P_{\infty,high-torque} \text{ (not optimistic)}$$
Computation Time = 2.1s

Figure 5.20 Comparison of node expansion performed by SBMPO during energy efficient planning in the multi-terrain environment of Fig. 5.19. The results compare planning with optimistic vs. non-optimistic energy heuristics. The color numbers are associated with the order in which the nodes are expanded. The first node expanded from the queue is assigned the value 0, the next node expanded is assigned value 1, and so on.

5.2.3 Future Research Directions in Learning Heuristics

Section 5.2.1 presented a neural-network-based approach to learning heuristics. This method involved extensive off-line simulation using randomized optimization sets. It seems possible to more efficiently choose the training sets to obtain more uniform improvement in the computational improvements achievable with the estimated heuristic. For example, in Fig. 5.16(b), some of the points appear in the undesirable black (upper diagonal) region of the graph. Making the choice of data sets more systematic is an area that requires further research.

Additionally, the choice of learning method can also be further explored. The neural network architecture used in the double pendulum example of Section 5.2.1 was not necessarily the best choice. Better and faster estimation and robustness with respect to data sets could possibly be achieved by using a different choice of neural network architecture and training. For example, deep learning for this problem has not been explored.

It should not be ignored that the learning method of Section 5.2.1 does not ensure optimistic heuristics. In fact, since optimistic heuristics are lower bounds on all possible values of the cost-to-goal, it seems impossible to guarantee that one can learn a nontrivial optimistic heuristic. In light of this, perhaps planning algorithm theory should be extended to discuss how the bounds of suboptimality and computational cost relate to the degree of non-optimism in the heuristic. One way of exploring this further might be to use k-fold cross validation to estimate the non-optimism of the learned heuristic.

Section 5.2.2 shows that there is a need for a multiple heuristic SBMPO framework, which can use either analytically derived or learned heuristics. The foundation for this is the work on MHA* [27]. As discussed in Section 2.5, the SBMPO framework can actually incorporate almost any planner. Hence, further research should develop an SBMPO/MHA* algorithm.

Application of SBMPO/MHA* to outdoor scenarios involving energy efficient planning in multiple-terrain and multiple-slope environments is an important one as this important planning problem has not been solved. As mentioned in Section 5.2.2, multiple heuristics are also needed when the robot itself has multiple dynamic models as is the case for legged robots operating in different gaits. To see this note that each gait requires different control system settings, and the control system actually changes the dynamics of the robot. This problem becomes even more complex when the legged robots are asked to move energy efficiently in multiple-terrain, multiple-slope environments.

Bibliography

[1] C. Ordonez, N. Gupta, B. Reese, N. Seegmiller, A. Kelly, and E. Collins, "Learning of skid-steered kinematic and dynamic models for motion planning," *Robotics and Autonomous Systems*, vol. 95, pp. 207–221, 2017.

[2] J. Y. Wong, *Theory of Ground Vehicles*. John Wiley & Sons, 2008.

[3] N. Seegmiller, F. Rogers-Marcovitz, G. Miller, and A. Kelly, "Vehicle model identification by integrated prediction error minimization," *The International Journal of Robotics Research*, vol. 32, no. 8, pp. 912–931, 2013.

[4] N. Sundararajan, P. Saratchandran, and Y. W. Lu, *Radial basis function neural networks with sequential learning: MRAN and its applications*, vol. 11. World Scientific, 1999.

[5] W. Yu, O. Chuy, E. G. Collins, and P. Hollis, "Analysis and experimental verification for dynamic modeling of a skid-steered wheeled vehicle," *IEEE Transactions on Robotics*, vol. 26, pp. 340–353, 2010.

[6] C. Ordonez, N. Gupta, W. Yu, O. Chuy, and E. G. Collins, "Modeling of skid-steered wheeled robotic vehicles on sloped terrains," in *ASME 2012 5th Annual Dynamic Systems and Control Conference Joint with the JSME 2012 11th Motion and Vibration Conference*, pp. 91–99, American Society of Mechanical Engineers Digital Collection, 2012.

[7] S. H. Young and D. G. Patel, "Robotics collaborative technology alliance (RCTA) program overview," in *Unmanned Systems Technology XX*, vol. 10640, p. 106400D, International Society for Optics and Photonics, 2018.

[8] J. Norby and A. M. Johnson, "Fast global motion planning for dynamic legged robots," in *IEEE/RSJ International Conference on Intelligent Robots and Systems (IROS)*, (virtual), 2020.

[9] P. Fankhauser, M. Bjelonic, C. D. Bellicoso, T. Miki, and M. Hutter, "Robust rough-terrain locomotion with a quadrupedal robot," in *IEEE International Conference on Robotics and Automation (ICRA)*, pp. 5761–5768, IEEE, 2018.

[10] M. Harper, J. Pace, N. Gupta, C. Ordonez, and E. G. Collins Jr, "Kinematic modeling of a RHex-type robot using a neural network," in *Unmanned Systems Technology XIX*, vol. 10195, p. 1019507, International Society for Optics and Photonics, 2017.

[11] M. Harper, J. Nicholson, E. Collins, J. Pusey, and J. Clark, "Energy efficient navigation for running legged robots," in *International Conference on Robotics and Automation (ICRA)*, pp. 6770–6776, 2019.

[12] J. M. Brown, M. P. Austin, B. Kanwar, T. E. Jonas, and J. E. Clark, "Maneuverability in dynamic vertical climbing," in *IEEE/RSJ International Conference on Intelligent Robots and Systems (IROS)*, IEEE, 2018.

[13] M. P. Austin, M. Y. Harper, J. M. Brown, E. G. Collins, and J. E. Clark, "Navigation for legged mobility: Dynamic climbing," *IEEE Transactions on Robotics*, 2019.

[14] T. Bretl, S. Rock, J.-C. Latombe, B. Kennedy, and H. Aghazarian, "Free-climbing with a multi-use robot," in *Experimental Robotics IX*, pp. 449–458, Springer, 2006.

[15] M. P. Murphy, C. Kute, Y. Mengüç, and M. Sitti, "Waalbot II: Adhesion recovery and improved performance of a climbing robot using fibrillar adhesives," *The International Journal of Robotics Research*, vol. 30, no. 1, pp. 118–133, 2011.

[16] T. Bretl, "Motion planning of multi-limbed robots subject to equilibrium constraints: The free-climbing robot problem," *The International Journal of Robotics Research*, vol. 25, no. 4, pp. 317–342, 2006.

[17] T. Chen and C. Guestrin, "XGBoost: A scalable tree boosting system," in *Proceedings of the 22nd ACM SIGKDD International Conference on Knowledge Discovery and Data Mining*, KDD '16, (New York, NY, USA), pp. 785–794, ACM, 2016.

[18] G. Ke, Q. Meng, T. Finley, T. Wang, W. Chen, W. Ma, Q. Ye, and T.-Y. Liu, "LightGBM: A highly efficient gradient boosting decision tree," *Advances in Neural Information Processing Systems*, vol. 30, pp. 3146–3154, 2017.

[19] S. Hochreiter and J. Schmidhuber, "Long short-term memory," *Neural Computation*, vol. 9, no. 8, pp. 1735–1780, 1997.

[20] F. Pedregosa, G. Varoquaux, A. Gramfort, V. Michel, B. Thirion, O. Grisel, M. Blondel, P. Prettenhofer, R. Weiss, V. Dubourg, J. Vanderplas, A. Passos, D. Cournapeau, M. Brucher, M. Perrot, and E. Duchesnay, "Scikit-learn: Machine learning in Python," *Journal of Machine Learning Research*, vol. 12, pp. 2825–2830, 2011.

[21] M. Y. Harper, C. Ordonez, E. G. Collins, and G. Erlebacher, "Parallel approach to motion planning in uncertain environments," in *Unmanned Systems Technology XX*, vol. 10640, p. 106400H, International Society for Optics and Photonics, 2018.

[22] R. Natarajan, M. S. Saleem, S. Aine, M. Likhachev, and H. Choset, "A-MHA*: Anytime multi-heuristic a," in *12th Annual Symposium on Combinatorial Search*, 2019.

[23] A. Majumdar and R. Tedrake, "Funnel libraries for real-time robust feedback motion planning," *The International Journal of Robotics Research*, vol. 36, no. 8, pp. 947–982, 2017.

[24] J. M. Maciejowski, *Predictive control with constraints*. Pearson Education, 2002.

[25] B. M. Reese and E. G. Collins, "Learning a cost-to-goal estimate for fast model predictive optimization based on graph search," in *American Control Conference (ACC)*, pp. 5038–5044, IEEE, 2017.

[26] J. J. Moré, "The levenberg-marquardt algorithm: implementation and theory," in *Numerical Analysis*, pp. 105–116, Springer, 1978.

[27] D. Youakim, P. Cieslak, A. Dornbush, A. Palomer, P. Ridao, and M. Likhachev, "Multirepresentation, multiheuristic A* search-based motion planning for a free-floating underwater vehicle-manipulator system in unknown environment," *Journal of Field Robotics*, vol. 37, no. 6, pp. 925–950, 2020.

CHAPTER 6

Contributions and Future Work

This book described Sampling-Based Model Predictive Optimization (SBMPO), a new paradigm in optimization-based kinodynamic planning that is unique in that it is actually based on an A* algorithm and relies on the use of A* heuristics. The underlying cost function can be a physically motivated cost function such as distance, time, or energy. The planned trajectories are able to respect the system dynamics and actuator constraints, and the dynamic models used by SBMPO can be linear or nonlinear, physics-based or data-driven (e.g., neural networks). Finding heuristics that work within this paradigm is non-trivial. For a single-input, single-output double integrator system, Chapter 3 develops heuristics for the case of time and distance cost functions. Chapter 4 then describes how these heuristics can be used in planning for much more complex systems. Links to computer code with examples is provided in the Appendix.

Several applications of SBMPO to planning using dynamic models were provided in Chapter 4, including momentum-based planning for AGVs and manipulators, spacecraft rendezvous maneuvers, energy efficient planning for skid-steered vehicles, and thermally informed motion planning for legged robots. Each of the applications was described in terms of

                    DOI: 10.1201/9781003623830-6

cost functions, associated models, formulation of an optimistic heuristic, and steps of the SBMPO algorithm specialized to the application. In most cases single or double integrator models were used as the propagation model. For momentum-based planning, the full dynamic model was used in conjunction with a computed torque controller to enforce torque constraints. For energy efficient planning a steady-state dynamic model is used. Hence, this book provides ways in which full dynamic models can be used in planning without using these complex models as the propagation model.

The importance of learning for trajectory planning with SBMPO was described and illustrated in the context of vehicle propagation models for wheeled and legged platforms in Chapter 5. It shows how a physics-based model can be augmented with data-based dynamics and how a data-based model can be constructed using neural networks. These models are used to plan for complex physical systems. As heuristics are key to fast computations in SBMPO and may be difficult to develop for many systems, Chapter 5 further motivates the need for learning heuristics, including the development of an SBMPO algorithm based on MHA*, i.e., multi-heuristic A*. It also presents early results on using massive simulation in conjunction with a neural network to learn a heuristic function.

Chapter 5 describes future research directions in learning models (Section 5.1) and in learning heuristics (Section 5.2). This research is vital for non-fragile applications of kinodynamic planning. Additional areas of future work can be in the development of heuristics for cost functions not yet considered in the SBMPO paradigm such as stealth or risk. Additionally, this planning paradigm can be applied to a variety of robotic platforms.

Code Repository

Source code to help users apply SBMPO to their system of interest is provided at `https://www.github.com/MobileRoboticsLab/SBMPO.git`

The repository contains a standalone implementation of SBMPO that aims to keep the code simple and easy to modify. Documentation guiding the user to the specific sections of the code that would need to be modified to fit their needs is provided.

In addition, the repository presents examples on how to integrate SBMPO with ROS. All code in the repository is continuously upgraded to reflect new developments in the SBMPO framework and auxiliary software components such as ROS, mapping libraries, and new software development practices.

A.1 UNICYCLE STEERING

The repository describes the implementation of minimum distance planning with a velocity-driven unicycle model operating in an obstacle field. The code includes scripts to visualize the generated results.

 DOI: 10.1201/9781003623830-A

A.2 ACKERMANN STEERING

The repository describes-implementation of planning with an Ackermann-steered vehicle using a minimum distance cost. The code includes scripts to visualize the generated results.

A.3 DOUBLE INTEGRATOR

Example code demonstrates the utilization of the minimum time heuristic with a double integrator is provided. The code includes scripts to visualize the generated results.

Index

Note: – Page references in *Italics* refer to figures and **bold** refer to tables.

For Product Safety Concerns and Information please contact our EU
representative GPSR@taylorandfrancis.com
Taylor & Francis Verlag GmbH, Kaufingerstraße 24, 80331 München, Germany

www.ingramcontent.com/pod-product-compliance
Lightning Source LLC
Chambersburg PA
CBHW070708190326
41458CB00004B/892